The

BODY BOOK

The
BODY BOOK

Feed, Move, Understand and Love Your Amazing Body

CAMERON DIAZ

with SANDRA BARK

HarperCollins*Publishers*

HarperCollins*Publishers*
77–85 Fulham Palace Road,
Hammersmith, London W6 8JB
www.harpercollins.co.uk

First published by HarperCollins*Publishers* 2014

10 9 8 7 6 5 4 3 2

Design © Headcase Design www.headcasedesign.com
Illustrations © Patrick Morgan
Cover reverses: Women who shared their images and stories for *The Body Book*,
photographed by Cameron Diaz.

Pages 114–15: classic computer: Anteromite/Shutterstock, Inc.; retro percolator:
Frank Anusewicz-Gallery/Shutterstock, Inc.; MP3 player: Tetat thailert/
Shutterstock, Inc.; vintage analog recorder: Valentin Agapov/Shutterstock, Inc.;
CDs and DVDs: Sarah2/Shutterstock, Inc.; VHS: Mathieu Viennet/Shutterstock,
Inc. Clock and calendar graphics on pages 118–19 by Headcase Design.

Cameron Diaz asserts the moral right to be identified as the author of this work.

A catalogue record of this book is available from the British Library.

ISBN 978-0-00-752205-7

Printed and bound in Spain by Graficas Estella

MIX
Paper from
responsible sources
FSC C007454

FSC™ is a non-profit international organisation established to promote the
responsible management of the world's forests. Products carrying the FSC
label are independently certified to assure consumers that they come from
forests that are managed to meet the social, economic and ecological needs
of present and future generations, and other controlled sources.

Find out more about HarperCollins and the environment at
www.harpercollins.co.uk/green

Dedicated to your body.

CONTENTS

INTRODUCTION: *Knowledge Is Power*

1

PART ONE   NUTRITION: *Love Your Hunger*

PART TWO   FITNESS: *The Body Wants to Be Strong*

PART THREE   MIND: *You've Got This*

CONCLUSION: *Now You Really Got This*

257

KNOWLEDGE IS POWER

———

Hello, Lady!

Thank you for picking up this book. Before we launch into Chapter 1, I want to tell you why I wrote it, what it means to me and what I hope it will mean to you.

As human beings, when we have experiences that bring us great joy and fulfilment, all we want is to share that excitement with others. Like when you eat something incredibly delicious and immediately turn to the person next to you and say, "Taste this!" Or when you hear a really amazing song and then download it for a friend because you can't wait for her to hear. Or when you learn a piece of information that just really blows your mind, something that feels so big and important that you want to share it with everyone you can find!

That's how I feel about this book. Everything in these pages is information that I use to live my life, information that makes me feel so excited and joyous that I just had to share it with you. That's why I've written it – because educating yourself about your body is one of the most important things that you can do. As you read through these pages, you're going to learn about nutrition and how to feed yourself deliciously and nutritiously. You're going to learn about fitness and how your body is affected by movement. You're going to learn about your mind, so that you can become self-aware and find your inner discipline. Because *nutrition* and *fitness* and *awareness* and *discipline* are not just words: they are tools. They are power. They are ways to care for yourself that empower you to be stronger, smarter, more confident and truer to yourself.

In the subtitle of this book are the words 'Your Amazing Body'. I believe just that: your body is amazing. Right now, whatever shape you are in, your body is an amazing machine that does so many cool things, from using the air

Educating yourself about your body is one of the most important things that you can do. Because *nutrition* and *fitness* and *awareness* and *discipline* are not just words: they are tools.

in our environment to keep your brain alive to turning a bowl of cereal into an explosion of energy that allows you to run down the street to catch the bus. And knowing how to take care of that body is the most important information you can ever learn. *Ever.*

Because your amazing body is the only body that you will ever have. The same body you've had since you were a baby is the body you will inhabit when you are seventy-five years old. And along the way, yes, it has changed, and it will continue to change – but it is still yours. No matter what shape it is, how much you love or hate it, whether it feels tired and worn down or lively and invigorated, your body is the most precious thing you have.

Your body is your past, present and future. It carries the memory of your ancestors, because you are made up of the genes given to you by your parents and their parents before them. It is the culmination of everything you have ever eaten, all the physical activity that you did or didn't do, all the efforts that you've made to understand and take care of it. And how well you care for it will determine how well you are able to live your life. So whether you wish you had longer legs or smaller hips, bigger boobs or less pointy ears, this book is for you. It's a guide to accepting what you have and *loving* it with everything you've got, to appreciating how incredible this physical vehicle is. It's a guide to maximizing your strength and endurance so that your body can take you everywhere you want to go in life: to all of your success, to the love of your life, to your passions and adventures. The body that you have is the body that will literally carry you there. So if you want to get to all of those wonderful destinations, you have to build the strongest, most capable, most powerful body that you can. You must learn to live well in *your* uniquely beautiful body.

But you can't do that if you don't know how. Unfortunately, as women, we are constantly being pressured about being more beautiful, about being thinner, about looking younger, or sexier, or blonder or more brunette. As women in today's society, we are encouraged to compare ourselves to other women when what we need to do is focus on our own strengths, our own capabilities, our own beauty.

That's why I wrote this book: so that together, we can learn the science behind the words and have the power that comes with knowing the truth about our bodies instead of absorbing the bias and misinformation that surround us. I'm not a scientist. I'm not a doctor. What I am is a woman who has spent the past fifteen years learning about what my body is capable of, and it has been the most rewarding experience of my life. Everything that I have, everything that I am, relates to my knowledge about my body. And I want you to have the same thing. I want you to know yourself, your power. I want you to be the most powerful, capable, confident woman that you can be. I want you to know what it feels like to have a relationship with your body, to feel connected to it. I want you to know the true joy of living in a body that is yours and yours alone, of knowing how incredible it feels to nourish yourself with good food, to move and sweat, to truly care for your health. Because once you posses that knowledge and are living in the body that you are meant to be in, you will find that your energy is boundless and that you can see and experience the world in a way that you couldn't before. And you will start to be able to use your energy in ways that you'd never even thought of before, just because you were too focused on what you thought you didn't have or couldn't do.

Because I want all of this for YOU, I spoke to people who are experts in medicine, nutrition, fitness, science, health and psychology, people who have dedicated their lives and careers to understanding and helping the human body and mind be the best that they can be. It is their knowledge that I share with you, and it is my own journey of learning and taking care of my body that I offer to you as an example of how to apply that knowledge so you can enjoy the benefits.

When you have finished reading this book, when you have integrated this information, when it is really inside of you, in your body, in your habits, you won't have to think about it anymore. It will become part of you. It will become *you*. When that happens, all of your energy is transformed into positive energy aimed at doing, accomplishing, being, creating things in the world, instead of worrying about how you look, why you feel tired, why those pounds just hang on and slow you down. Just think of all the amazing things that you could create in the world if you felt free, powerful, self-confident!

But that transformation doesn't happen in a day, in a read, in a hope. The truth is that there is NO SUCH THING as a quick fix or a magic pill that will make you a healthy, happy human being overnight. To be healthy means that you must not only learn how your body works and what it needs to be healthy, but you must also apply that knowledge as consistently as you can towards making the best decisions you can to achieve that health. It's an ongoing endeavour, not a one-off deal. That is why knowledge is so powerful – because when you really know something inside and out, you can see all of the opportunities to apply that knowledge through action every single day.

Here's what this book is not: It is not a diet book. It is not a workout regimen. It is not a manual to becoming a different person.

Here's what it is: a guide to *becoming yourself*. Because as you begin to learn more and more about your body, something amazing starts to happen: you begin to transform, on the inside and outside. You start to see how being healthy brings happiness into your life, how good it feels to be strong and able, how feeling good on the inside has an effect on everything else in your life. You will be the most beautiful and healthy and confident woman that *you* can be, and you deserve that, because YOU ARE MORE BEAUTIFUL THAN YOU HAVE EVER IMAGINED.

By the time you have finished this book, you will be informed about how your body works on a basic level. You will be aware of how your mind and body work together. You will be more powerful, because you will understand how amazing, how incredible, how beautiful your body already is, right now, right this very second.

So what I'd love for you to do is to use this book as a guide to understanding that amazing body and to help it to become what it has always been meant to be. Make this book your own! Get a pen and write in the margins. Take notes. Bend the corners of the pages. Ask questions. Seek answers. And get ready to meet your true, powerful, healthy, confident, amazing self.

PART ONE

NUTRITION

Love Your Hunger

CHAPTER 1

YOU ARE WHAT YOU EAT

ONCE UPON A TIME, you were so small that you weren't even visible to the human eye. You were just a cell in your mother's womb, a single microscopic speck. And then you became two cells . . . then four cells . . . then eight cells, and those cells kept multiplying, copying and differentiating until you became a hundred trillion cells, each with a distinct purpose: brain cells and skin cells and heart cells and stomach cells and blood cells, cells that produce tears and cells that produce milk and cells that make you sweat and cells that grow hair and cells that help you see.

The hand that is holding this book began with that tiny cluster of cells. Your entire body began as a barely perceptible dot, and somehow, you grew into this amazing and glorious specimen. *How does that happen?* How did you develop from a tiny spot of life into this incredible living, breathing, running, laughing being? How did your bones and muscles grow into what they are today? What about organs like your brain and your skin, or the most important muscle in your body, your beating heart? What makes them continue to grow and function, and how do they come to be healthy or sick, strong or weak?

There's one word that can answer all of these questions, and that word is NUTRITION. The nutrients in the food that you eat determine the way your cells develop, grow and thrive (or not). When you were a bun in your mother's oven, your development was, at least partially, a result of her lifestyle and the nutrition that she put into her body (the other part was genetics, which she couldn't exactly control). And now as a grown human made up of trillions of cells, your health depends on the nutrition you offer your body every time you eat.

EXCUSE ME ... WHAT'S A CELL?

When I began writing this book and I was getting all CSI about the body, that was one of the questions I asked. What is a cell, exactly? One of the great things about that question is that humans have only known that cells exist for around three hundred and fifty years. Before 1676, nobody had a clue about cells, because nobody had ever seen them. Then a guy named Antoni van Leeuwenhoek peered at a slice of animal tissue through a stronger-than-ever-before microscope, and to his great surprise, he discovered that living things were actually composed of tiny little 'rooms', which he called cells.

Three centuries later, we know that the human cell is a complex, living structure made of fat and protein (which are, not coincidentally, two key components of your nutrition). After you eat and your digestive system has its way with your food, your cells get to have a go. Your cells are basically miniature processing plants that use oxygen to convert nutrients from the food you eat into the energy that you use.

Your cells are busy little bees. Some of them are the red blood cells that make you red-blooded. Some are osteoblasts, which are the cells that make up your bones. And all of your cells hold your genes, in the form of DNA. That means that everything about you – from the colour of your hair and your eyes to your blood type to your risk for developing certain diseases – lives in your cells, including your ovarian cells, which will create the egg cells that are a collection of your genes and can be passed down to the next generation.

Your many different kinds of cells work together as a team to create your physical being – and whenever a member of the team can't function at its best, it's you who winds up in the doctor's surgery. Which is why you must act as a nutrition delivery system for your cells, finding and consuming the most nutrient-rich foods possible, so your cells can do everything they want to do: protect you, energize you, heal you, as well as enable you to keep thinking and breathing. (Thank you, brain cells and lung cells.)

Because *you are what you eat.*

YOU ARE WHAT YOU EAT

How old were you the first time you heard that saying? I've been hearing it since I was a little kid, but it wasn't until I was an adult that I really understood what it meant. When I was young, it just sounded like any other expression adults said – not like a piece of wisdom I could apply to my life. Back then, I didn't know how to connect the dots. I didn't know that the food I ate had anything to do with how I felt, let alone that it literally powered the cells that powered me.

Today, I know better, and I know that this is what it means: that the food we consume over the course of a day creates the experiences we have in that day. Because what we eat carries the stuff of life – our lives.

Your days may be full of energy and clear thoughts, happiness and gratitude, productivity and advancement, or they can be the complete opposite. Sluggishness, foggy thoughts, sadness, regret . . . basically, a bad day full of wasted opportunities. It took me a long time to really understand that, but I finally get it: If I *eat* rubbish, I'm going to *feel* rubbish. If I eat good healthy food full of energy, I'm going to be full of energy.

Today, tomorrow and twenty years from now, your nutrition is worth your attention and your time, because nutrition is health, and health is everything.

The food we consume over the course of a day creates the experiences we have in that day. Because what we eat carries the stuff of life – our lives.

WHAT DOES IT MEAN TO BE HEALTHY?

The word *health* gets thrown around a lot these days, so let's take a moment to clarify what I mean when I talk about being healthy. When I refer to health, I'm talking about having a body that is working at its optimum, a body that has the energy to go all day without crashing, a body that can fight off illness and keep you strong. I'm talking about feeling amazing in your skin, in a body that can wake up in the morning, get out of bed, make breakfast and get moving. I'm talking about having a mind that can be clear and productive, thoughtful and happy.

If you're healthy, you are incredibly lucky, and you must do whatever you can to preserve that health. If you aren't, you must do whatever you can to take care of yourself, to offer reinforcements to your immune system, to give yourself every ingredient your cells need to help your body function as well as possible so that you can feel as good as possible.

When I think about how debilitating it is to just have a common cold, I can't imagine what it must feel like to have an illness that is life altering or life threatening. When my body doesn't work the way I want it to, when I can't spend time with friends and family because my body hurts every time I move, I HATE it. Even though I know that in a few days I'll feel better, it's still frustrating. That's the kind of feeling that makes me want to do whatever I can to keep my body healthy.

No matter where you're starting from, one of the most important things you can do for yourself is to love your hunger: to eat for nutrition and give your body and every little cell that it contains exactly what it needs to help you thrive.

And that includes your taste-bud cells.

CHAPTER 2

FOOD, GLORIOUS FOOD!

LOVE FOOD. I LOVE to cook it, love to eat it. Love to cook for myself, love to cook for others, love when friends and family cook for me. I am always sharing delicious meals with the people I care about. We bring each other meals, we invite each other over for dinner, we deliver the shopping when someone isn't feeling well. I'll invite my friends to my house and we'll all cook together, each of us making the foods we love, because everybody has their speciality. One of my favourite memories is hosting a huge Cuban-themed dinner party at my house last Christmas. My mum and I spent an entire day cooking our traditional Christmas meal: roast pork, roast chicken, black beans and rice, avocado salad . . . it's a meal that's created with lots of love. We invite our friends and their families, set out a huge table, and all of the kids just run around on the grass in between bites while the adults tuck into the feast.

I've always enjoyed the emotional charge that we get from sustaining one another with delicious meals. There's something so warm and nurturing about being cooked for, and something so happy and rewarding about cooking for others. When I was young, every day after school, when my mum got home from work, we would catch up in the kitchen while we cooked dinner together. That dinner provided physical nourishment for our family, but it also gave me emotional nourishment.

Food is everywhere in our lives. Eating can be about celebrating culture, tradition or religion. Human beings eat at weddings and at funerals, at potlucks and at fancy dinner parties. We eat on dates. We eat at business lunches. Food is part of family life, like when we cook big holiday feasts. It's part of our social fabric, like when we meet friends for dinner after work. And meal by meal, it's what's *on* our plates that will define our health.

If we want to have good health, we must eat good, real, whole food. And if there's one thing I love to love, it's good food – I am the kind of girl who will literally lick the plate. The great thing is, we can eat for nutrition *and* enjoyment. You can eat food with flavours *you* crave while also providing yourself with the nutrients *your body* craves.

Real food. Good food. Delicious food. Chewy crunchy spicy glorious food. Whole and hearty food that gives our bodies the building blocks of our lives, health, energy, and vitality.

FAST FOOD ISN'T REALLY FOOD

When I talk about eating for nutrition and eating *good food*, eating *real food*, and eating *whole food*, I'm talking about eating foods that grow in the earth or are sustained by the earth and that have not been tampered with by technology.

How? By avoiding fast food and processed food. By choosing whole grains, vegetables, and fruits that are as close as possible to their state when they came out of the ground. Fast food and processed food may have started out as food, but by the time you encounter them, they have been saturated with preservatives, painted neon with artificial colours and doused with fake flavours so much that they aren't actually food. Seriously. I don't even see these 'food products' as food because *they do not give me health*. They don't give you health, either. In fact, as you'll learn, they don't even really satisfy your hunger.

As you read this book, we're going to look at why modern inventions like processed snack foods and fast foods are not a good source of nutrition and what the impacts of living without nutrition are for your health (and your life).

Believe me – I know about fast food. I grew up on fast food. My mum cooked every night, and we ate dinner at home, but out and about as a teenager, I was a bit of a fast-food junkie. My friends and I would roll through the drive-thru, and I would get a double cheeseburger with french fries *and* onion rings. When I was in junior high, my friend's brother worked at Taco Bell. Every day after school, I would come by and order one bean burrito with extra cheese and extra sauce, no onions, and he would always give me two. Every day after school, I ate two bean burritos and a Coke. Every single day, for three years, that was what I ate. Every single day.

If you are what you eat, I was a bean burrito with extra cheese and extra sauce, no onions.

And all the while I was eating burgers and burritos and onion rings and french fries and sodas, I had the worst skin. I mean, I had terrible, terrible skin. It was embarrassing, and I did everything I could think of to make it go away. I tried to cover it with makeup. I tried to get rid of it with medication – oral, topical, even the harshest prescriptions. Nothing helped for very long.

The spots were still there through high school and into my twenties, while I was modelling and acting. It was really challenging to cover them up for the cameras; it was awkward and embarrassing and frustrating, and I always felt really bad about myself. But I kept eating that fast food, still following habits that formed at a young age when I had no idea that food could affect my strength, my energy, my ability to function – or my skin. I never even considered that there might be a connection between my meals and the way I felt or the way my skin looked. And it was just so easy to go through the drive-thru and get my Southwestern chicken with cheese and bacon, and of course, my onion rings and fries, with a side of ranch dressing.

I was there so often that they knew me at the drive-thru.

I was always a skinny kid—a skinny adolescent, a skinny teenager, who grew into a skinny adult. All I had ever heard from people about the way I ate was, "You're so lucky! You can eat whatever you want and you stay thin!" I wasn't gaining weight, and I didn't have a microscope to see inside my body and look at my unhappy cells . . . so I just never thought that my diet could be the root of my problems. But the truth is that everything we put into our bodies affects us, no matter what body type we have. Some things affect us positively, like whole foods that give us nourishment and fuel to do the things we love. Some things affect us negatively, like processed foods that are empty of nutrients and full of chemicals, artificial colours, and preservatives that can disrupt our hormones and prevent our bodies from functioning the way they were born to. It's just that simple.

My issues with my skin persisted until my late twenties, when I started cooking for myself again and stopped eating so much fast food. As my eating style evolved and I quit putting processed foods into my body, a funny thing happened . . . my skin began to clear up. My acne wasn't totally gone, but it was significantly better. Looking back, I realize that I hadn't needed those prescription drugs, those vials of potions and creams. I hadn't needed to be angry at my skin or feel bad about myself. I had just needed to LISTEN TO MY BODY. I may not have had a microscope, but that acne was my body's alarm system, its way of telling me, "Stop! Give me what I need so that I can do the job I'm supposed to do!" As I began eating whole foods and cutting out the salty and sugary and fried fast foods, my system gradually started to find its balance and my skin began to clear. It's definitely possible that hormonal changes and other factors may have played a role in my acne, but it's also definitely true that I saw a dramatic change in my skin when I changed my diet. And as I made more changes, I started noticing other responses in my body to the foods I ate – like how my stomach felt or how bloated I would get after a meal. I began to realize that by adjusting my food intake I could affect not only my skin but also my energy levels and my stomachache levels . . . not just the way I looked but the way I FELT. If you're anything like me you may have had the experience of wondering why you don't feel at home in your body, or the feeling that you're not living in your true body. Well, if you've been eating processed foods like I had been, you're *not* in your true body – but you can be.

Experiences like these were the real beginning of my journey with food and nutrition. When I began to realize that the answer to my problems wasn't in a beauty cream or the medicine cabinet, I wanted to know more. I had some friends who were into nutrition, and I asked them to share their tips with me. The more I learnt about how much food affects me, the more questions I had, so I started reading and listening to programmes. The more answers I got, the more I wanted to know. Everything I learnt was an invitation to understand a little bit more. And I am still seeking, listening and learning today. I know that curiosity counts, that taking an interest, asking questions and following through on what I learn leads to results.

Now that I understand that I create the experience of my entire life by what I eat, I have been transformed. So if there's an issue that you're grappling with, whether it's your skin, your weight, your heartburn or your mood, instead of adding layers of pills and creams or other quick fixes, start at the foundation: your nutrition. I swear to you that the food you eat has a huge impact on the person that you are, mentally and physically, in the next few hours and for the rest of your life. My experience as a human being changed when I realized that I am what I eat and when I decided to give myself the best chance possible to feel as good as possible.

When we're young, it's our parents' job to make sure we stay healthy – that we get a good night's sleep, eat breakfast and go to school with a packed lunch or lunch money. Somehow, we can lose sight of these basic pillars of health after we grow up, and the habits we need to be happy, vibrant people can take a backseat to the busy-busy of workschoolfamilyfriendshobbies.

The responsibility to be healthy is in your hands – no one else is going to do it for you. So ask yourself: do you want to live in a body that allows you to do the things you want to do, a body that is full of health and capability, that you are proud to call your own? It's your choice.

And the amazing thing is that you don't have to choose between health and good food, because you can eat food that is good for you and also tastes great. You can have deliciousness and healthiness. Real food is everything. It's pleasure. It's fuel. It's nutrition. It's family. It's life.

Now that I understand that I create the experience of my entire life by what I eat, I have been transformed.

CHAPTER 3

HUNTER, GATHERER, DRIVE-THRU-ER

YOUR BODY IS A gorgeously designed machine. Just like the other machines that power your life, your body needs fuel – and not just any form of fuel will do. If the red light on your dashboard lit up to warn you that your car was running low on petrol, you wouldn't buy a litre of tomato juice and pour it into the engine, would you? Of course you wouldn't – that would be ridiculous. Cars don't run on tomato juice. They run on petrol, diesel and electricity. Your cells also run on fuel, and just like your car, it's important to give your cells proper fuel for optimal performance. Whether you call it fuel or food or nutrients, your energy comes from what you eat. It makes everything you do, everything you think or say or feel or want, possible. Food keeps you *alive*.

Now, I can almost hear you saying, "No kidding, Cameron. Of course food keeps me alive. That's why I eat it." Well, yes, I know that you know that much . . . but do you understand the difference between whole foods, the ones that give you life, and processed foods, and the ones that have about as much nutrition as the plastic wrappers they come in? Do you know how your system extracts nutrients out of your food and how your body turns them into energy? Do you know what *glycogen* is and what its function is in your body? Do you know that *carbs* are the basis of your body's energy system? That most digestion takes place in the small intestine – not your stomach? That you need to eat the right amounts of the right kinds of fat if you want to stay healthy?

These may sound like issues that you don't need to concern yourself with. They may sound like too much science or too many instructions. But trust me: through the history of mankind, this information, how to feed ourselves to keep ourselves alive, has proved more useful to our species than the invention of gunpowder, rocket launchers or texting.

NUTRITION IS SURVIVAL

Thousands of years ago, human survival depended upon the bounty of nature and our ability to capture or kill the animals and harvest the plants that we happened to find. As hunter-gatherers, we needed to have a sophisticated understanding about which roots and berries were safe to eat and which ones were poisonous. We spent a lot of time figuring out how to track and kill beasts much more powerful than ourselves and probably even more time ensuring that we were always near a reliable source of water.

These may sound like issues that you don't need to concern yourself with . . . but trust me: throughout the history of mankind, this information has proved more useful than the invention of gunpowder, rocket launchers or texting.

Today, even though we live in a society where we hunt and gather burritos that need to be microwaved instead of water buffalo that need to be speared, we are still hunter-gatherers. The problem is that we are hunting and gathering processed food. Even though we are modern humans with central heating in our caves and a plethora of ready-to-eat meals at the deli counter, we are still humans. I might be able to call a taxi by waving my smartphone around, but I still have the same nutritional needs as those guys who made fire by rubbing two sticks together. We have the same need for energy, and the same internal response systems that make us want to chase food when we see it. And we have the same singular life goal: to find something to eat.

I mean, think about it – most of your life revolves around getting food. You may not be wielding a spear in the forest, but you learnt to use a fork as an infant so you could eat food. You learnt to speak so you could ask for food (one of the first words a baby says is usually "MORE!"). Then you learnt to do maths so that you could count money to buy food. You got an education so that you could get a job to make the money to provide the necessities of life, by which I mean food. Sure, having a roof over your head is part of it, but if you had enough money for only one thing, that one thing would be food. Everything you know and have learnt comes back to the most central thing in our lives: food.

Just because you can put something in your mouth, chew it, swallow it and then poop it out doesn't mean it's food. It just means you can chew it, swallow it and poop it out.

So it stands to reason that knowing which foods to eat and which foods to avoid should be as basic as knowing how to tie our shoes, brush our teeth and recite our ABC's. But shockingly, it's the thing that we seem to know the *least* about. Over the past few decades, with the advent of processed foods, human beings have begun to overeat these prepackaged fat-sugar-salt bombs instead of eating wholesome whole foods in healthy quantities. The awful result is that too many people spend every day eating food that causes them to feel nauseated, bloated and sluggish, food that causes their bodies to gain weight and their skin to break out, food that gives them headaches and heartburn, food that sends them running to the bathroom or prevents them needing it at all.

Even more awful is the fact that the cumulative result of all of these unhealthy meals is a society that has a growing population of sick and unhappy people who don't even know that it is the food they are eating that might be making them sick – that the food they are eating is not even food! Here's a secret: just because you can put something in your mouth, chew it, swallow it and then poop it out doesn't mean it's food. It just means you can chew it, swallow it and poop it out.

Just like when I had no idea why I felt like crap or why my skin kept breaking out for all those years, a lot of us just haven't learnt the basics of how to fuel our bodies. Until we take the time to learn about how we function as human animals, we will continue to make ourselves sick. And the scary part is, the kinds of 'sick' I'm talking about above (bloating, heartburn, skin problems, etc) are mild symptoms of much bigger problems – of the kinds of diseases that can kill us, and are killing us at alarming rates. This country is facing a crisis of obesity in both adults and children that has reached epidemic proportions.

In 2013, as I am writing this book, approximately one in three Americans is obese. That goes for kids, too: the Centers for Disease Control reports that more than one-third of American kids are overweight or obese. Adolescent obesity has *tripled* since the '80s. Small wonder that the American Medical Association recently announced that it is officially classifying obesity as a disease.

Obesity and its accompanying diseases are killers. Our generation is witnessing a profound shift in the way human beings live on this planet: for the first time in history, more people are dying of the issues that come along with an *excess* of food rather than a *shortage* of food.

Throughout human history, expected human life spans have slowly, steadily risen. If you were a twenty-year-old living in 1750, you might expect to live for another ten or twenty years. Today, a twenty-year-old woman can expect to live for another fifty-five years . . . if she is healthy. But obesity is threatening to shift the rise in life expectancy that has been slowly accruing and make it slide backwards. According to a 2005 study published in the *New England Journal of Medicine*, this is the first generation of American kids whose life expectancy is shorter than that of their parents. Got that? For the FIRST time in history (not counting wars and plagues), our life expectancies are getting SHORTER instead of LONGER. The quality and the quantity of the food we consume in the Western diet is causing us to literally eat ourselves to death. THAT'S CRAZY! Why are we using food to kill ourselves instead of enjoying it for what it is intended to do – to keep us alive and healthy?

Our world today is built around the central idea of convenience. You can go to the grocery store and buy the fruits and vegetables someone else has grown and harvested for you (and often shipped halfway around the world to you), the meat they have raised and slaughtered for you, and the bread that has been baked for you. You can also choose from thousands and thousands of packaged products that have been invented to meet your every possible craving and desire. But if we didn't live in a society where we could just pop by an Asda and pick up everything from pineapples to pork chops, we'd have to do all of the work of food production ourselves – growing, harvesting, slaughtering, cooking, baking. In fact, the pursuit of our dinner would be our entire lives, because there would be very little time to do anything else.

Because food is essential to life, you would seek out and eat whole foods full of all the energy and nutrients you needed. And once you found those foods, you would eat as much of them as you could. Your body would store that food in the form of fat so that when there wasn't enough food available, you would have a reserve of energy to fall back on. You would be grateful for the layer of fat padding your body because you knew that your body would *use* it to survive winters when nothing grew and to hunt and kill and gather more food when you could.

Back in those days, our ability to eat even when we were not hungry had the potential to save our lives. Then things changed. Instead of hunting and foraging, we gradually began to grow our own food and raise our own animals. And life got a little easier. But not much easier, because there was still a whole lot of work to be done.

If you were alive in the early days of agriculture, you had to figure out which crops would grow in your environment and how to make them flourish. You would spend all day in the field, up before the sun. You would take great care in cultivating the crops, and you'd tend your livestock properly. You would be sure that you were growing as many different plants as you could and that the animals you were raising for either milk or for meat were given food that would keep them healthy because, like people, animals are what they eat. You would have no other choice, because it takes a lot of work to create this necessity that we call food.

Eventually, all of that agriculture led to cities and towns where people could be postmen and poets in addition to hunters and farmers. Nowadays, the responsibility for feeding the human race is divided up among people who specialize in cultivating all of those different nutrients, thereby freeing up the rest of us to pursue pleasures and goals we would not have time for if we had to spear or grow our own food. That's the good part.

The bad part is that as we've created more convenience in our lives by giving other people responsibility for cultivating food for us, we've let go of the knowledge, concern, and responsibility for our own nutrition. We've outsourced it. Hippocrates, the father of Western medicine, famously said, 'Let food be thy medicine and medicine be thy food.' And he said this over 2,500 years ago. But somehow, over the centuries, as we've become smarter about things like plumbing, transport, and nail varnish technology, we became idiots about the very thing that keeps us alive.

THE PAST CENTURY IN EATING

A hundred years ago, the United States was not a place where you could expect to find tacos and Chinese food and pizza in every town (the first pizzeria in New York City opened around 1905). In fact, you probably wouldn't have been able to find any of them. What you would have found were a few small restaurants owned by locals, offering foods prepared in local ways with local ingredients. No fast-food chains, no food courts, no thirty-minute pizza delivery.

It was also just a century ago that technology allowed companies to begin to mass-manufacture foods. Around 1910, advertisers began to encourage housewives to stop baking their own bread and save time with store-bought loaves. In the 1930s, the modern kitchen, which let people store food for longer and cook food without as much hassle, made convenience even more of an option. That same decade, Kraft introduced instant, ready-mix macaroni cheese, and Nescafé came out with instant coffee. In the '50s, TV dinners replaced family meals around the table, and in the '60s, the same decade that put fondue on the map also introduced Weight Watchers and diet shakes. More convenience foods, less wholesome food, and suddenly everybody needs to go on a diet.

Over the next decades, Americans started eating more and more processed foods and fad-dieting like crazy. Sound familiar? Some of the details may be slightly different but the same thing happened in the UK. Thankfully, over the past few years, there have been some wise voices rising up to tell us the truth about how processed foods damage our health, and why whole foods are essential for our well-being.

1900s:

The first affordable car. Cadbury's Dairy Milk. Shorter skirts. Tuna in cans. Electricity in urban areas. Tea bags. First fish finger recipe published in 1900. By 1903 Huntley & Palmers produced over 400 different biscuit varieties.

1910s:

Cocktail parties. Suffragettes. Cadbury's Bourneville Plain Chocolate. Kit Kat. Campbell's promotes its soups as recipe ingredients.

1940s:

Rationing. Victory gardens. Spangles. Instant Whip. Ministry of Food produces posters, leaflets and Food Flashes, teaching housewives to be creative with rationed foods.

1950s:

Everyone in Britain gets an extra pound of sugar and 4 ounces of margarine to mark coronation of Elizabeth II. Barbecues. TV dinners. Cabbage-soup diet. Grapefruit diet. The first diet soft drink: No-Cal ginger ale. Walls doubles its ice cream production capacity.

1980s:

Fluorescent jeans. Acid-wash jeans. Portobello mushrooms. Gardenburgers. Red Bull. Wispa. Boost Coconut. Twirl. Beverly Hills diet. Oat bran. The Olive Garden. Fruit Roll-Ups. Healthy Choice frozen dinners.

1990s:

McDonald's McLean Deluxe. Fat-free Pringles. Fat-free ice cream. Artisan breads. Deep-fried Mars bars. Ben & Jerry's homemade ice cream appears in UK. Chicken Tikka Masala declared UK's national restaurant dish. Jamie Oliver reflects growing popularity of cooking as a leisure interest.

1920s:

Prohibition. Flappers. The 'cigarette' diet. Processed cheese. Ice lollies. Walls begins production of ice cream in 1922. Birds Eye frozen foods. Crunchie bar.

1930s:

The Depression. The modern kitchen, including refrigerator and stove, is introduced. Nescafé. Spam. Weetabix. Mars bar. Smarties. Quality Street. Maltesers. Huntley & Palmer use all types of available media to promote their products.

1960s:

Gloria Steinem. Martin Luther King. Sit-ins. Bell bottoms. Julia Child. Fondue. Casseroles. Single-serving ketchup packets. After Eight mints. Twix. Marathon. High-fructose corn syrup. Weight Watchers. Slimming World.

1970s:

Rubik's Cube. Estate cars. Screwball. Egg McMuffins. Perrier water. Recyclable soft drinks bottles. Yorkie bar. Double Decker. Many Cadbury brands see vast increase in sales including Flake, Dairy Milk, Whole Nut and Fruit and Nut.

2000s:

Heinz Funky Purple EZ Squirt (coloured ketchup in a tube). McDonald's premium salads. Farmer's markets. Milk-and-cereal bars. Pinkberry. Low-carb diets. Grass-fed beef, free-range chicken, organic produce and other specialist products revive interest in smaller-scale production.

2010s:

Kale. Quinoa. Food trucks. Vitamin Water Zero. Salads that contain more than a thousand calories served in restaurants. Horsemeat scandal: horsemeat found in frozen lasagne. Cake pops. Mini-cupcakes. Posting calories in fast-food outlets. Gluten-free diets. Veganism. Juicing.

As babies, we learnt to say, 'More', and we haven't stopped since. Look around you. We are inundated with food. Restaurants on every corner. Processed food on nearly every shelf in our grocery stores. Food at the petrol station. Food you can get in five minutes simply by driving up to a box and shouting into it. Energy bars that are substituted for real food, even though they very rarely contain as much good stuff as an apple. And then there are the advertisements everywhere for restaurants boasting of excessive portions of gooey, greasy, fatty dishes, and restaurants that offer two-for-one deals so you can eat out one night and take a second helping home for the next night (sounds like a good bargain until you consider how much you'll have to

> Without a real understanding of how our bodies work to process food and what real food really is, we are like townspeople living under a spell, cursed to repeatedly make bad choices and wonder why we don't get the results we want.

spend later to lose the weight and manage your health). We've come to believe that *more* is better when it comes to food – from bulk-buying at food warehouses to buy-one-get-one-free deals at the grocery store. It's understandable that we're attracted to more – after all, as hunter-gatherers, more was the goal. But now we live in a culture of excess, not deprivation. And all of that MORE is giving us LESS of something else: nourishment. We are living in bodies that are overfed and undernourished! That's right: you can eat an abundance of food and still be malnourished if the foods you are eating don't provide nutrients, or if they don't offer *nutrition*.

Without a real understanding of how our bodies work to process food and what real food really is, we are like townspeople living under a spell, cursed to repeatedly make bad choices and wonder why we don't get the results we want. Learning the right information is like having the spell lifted. You will understand why some food gives your body the energy to power through your day and some leaves you exhausted by lunch and wishing the day were over at three. You will learn how to make the best food choices you can for the time and money and environment you have.

If you want to feel *more* capable, *more* powerful and empowered, then you must take responsibility for learning about nutrition.

I know how hard it is to be constantly confronted by all of the greasy, delicious choices that are at our fingertips 24/7. When I get that scent in my nose, I'm like a hound dog – I want to find the source, push to the front of the queue, and place my order. But then I remind myself about how I'll feel after I eat that meal, and it's not a pretty picture. I want to feel good inside and outside, so the only MORE I truly want on a regular basis is MORE NUTRITION.

CHAPTER 4

HOW TO LOVE YOUR HUNGER

'M GOING TO LET you in on a little secret: humans get hungry. We spend the first few months of life crying to let people know that we're hungry so that someone will feed us. Then we fall asleep, wake up when we're hungry again, and cry some more. Babies don't cry because they're bored and want a snack or because crisps would make a really nice accompaniment to the rotations of the stars-and-moons mobile above the cot or because they're craving some Belgian chocolate ice cream. They cry because their cells need nutrients to keep growing and developing, and the uncomfortable feeling of hunger is how our bodies let us know that we need that nourishment.

Now you're perfectly capable of feeding yourself. And you have plenty of opportunities to do so, because humans get hungry five or six times a day, every day of every week of every month of every year. That adds up to more than two thousand times a year that you get hungry. That's a lot of hungry, and a lot of opportunities to help or harm yourself. So let's take a look at what hunger really is, why you get hungry, and why you should *never ignore your hunger*.

WHAT IS HUNGER?

As we discussed earlier, your body operates like an elegant machine – and just like any other machine, it needs fuel to keep running. Hunger is your body's warning that you're almost out of fuel. Just like your car has a red light on the dashboard to signal when it's close to empty, hunger is that red light for your body. If you ignore this warning or try to outsmart it, it doesn't go away – in fact, it intensifies. Ignore it long enough, and you'll find that you start to lose focus and lose steam. You become inattentive, irritable and not really able to do anything well. Finally, when you can't take it anymore, you'll eat any food you can find, no matter what it is, no matter how it tastes. And since your body was so desperate for fuel, you will probably eat a lot more than you really need, just because you didn't feed your hunger when your body first prompted you.

Here's what's happening inside you during this process: when your empty stomach begins to growl, your cells are depleted of the nutrients and energy they need to function, and your body puts in a request for a refill. This is an important thing to remember, because while hunger is 100 per cent healthy, some diet plans and magazines and occasionally even our own imaginations seem set on convincing us that hunger is something there to trip us up or trick us. Here's the fact: hunger is your body urging you to take care of yourself, to give it energy so that you can live your life.

Have you ever wondered why, even when you feed your hunger, you just get hungry again three hours later? It's very simple: everything that you do requires energy. You burn energy while you are sitting, standing, running, walking, talking. Just thinking requires an amazing amount of energy – 20 per cent of your body's energy supply is used by your big, gorgeous, energy-hungry brain. You even burn energy while you're sleeping, so your heart can keep beating and your blood can keep pumping. Energy powers your conscious actions, like pouring yourself a glass of water, and it powers your unconscious life-giving mechanisms, like breathing and temperature regulation. Every moment of your existence requires energy. This is why you get hungry. And this is why feeding your hunger is necessary for your existence.

Hunger is not your enemy. It is a signal from the deepest part of you, prompting you toward survival. That's why it's *so* important to listen to your hunger and to honour it by choosing the best possible source of energy that you can find. What an amazing revelation! *You never have to be hungry anymore.*

FEED YOUR HUNGER
FIRST THING IN THE MORNING

When I was a child, my mother would wake me up every morning in time to cook breakfast for myself. Make some eggs, pour some milk and cereal in a bowl, whatever I had time for – I could not leave the house without eating breakfast. As I got older, even if I wasn't always eating the best breakfast I could possibly choose, I still was not allowed to leave the house without eating something. Breakfast was just a part of my morning: regular, expected, normal. It was a habit.

Because that habit was just a natural part of my life, I never really thought about how essential it was to my well-being. It didn't occur to me that when I felt chirpy and bright-eyed and raring to go by the time I left the house, it was because my body had fuel. (Thanks, Mum!)

I moved out of my parents' house when I was seventeen. For the first time in my life, I had my own schedule. I had a job – I was a model – and I was living on my own, and what all of this meant was that my mum wasn't waking me up anymore. I was waking myself up. Adulthood!

Sort of. My morning routine was structured around maximizing the amount of time I spent in bed. Who cared about cornflakes when ten extra minutes curled up under the duvet were on the line? I would compute how long it took me to wake up, shower, and get out the door, because sleep was my priority. What could be more important than sleep?

Whether I was going to an audition or showing up for set, without that in-built maternal alarm rousing me from slumber with enough time to prepare and eat a satisfying breakfast, I developed a new habit: skipping breakfast. At the time, I didn't realize the price of those ten extra minutes of sleep. I didn't realize that the nutrition I was cutting out was essential for my mental and physical health. So here's what would happen: after a casting, I would wind up back at my modelling agency office, and I would just be spinning out of control. With my blood sugar running low, either I would feel anxious and worried or I would be yawning all over the place and barely able to hold a conversation. (How could I be so tired after my ten extra minutes of sleep?!)

And my agent would ask, "Have you eaten?"

And I'd say, "Um, no, I haven't."

What's so crazy is that it wasn't until she brought it to my attention that I would realize that the reason I had no energy and couldn't think straight, that the reason I felt out of control was because I was ignoring my hunger. So I'd go downstairs, and I'd order lunch: a huge platter of chicken, greens and some mashed potato. After two bites – *ahh*. I could think again. I could breathe again. I'd feel like a human being again.

It took ages before I really connected the dots and realized that the reason my mum made me eat before leaving for school every day was because she wanted me to have the fuel to get through the morning with flying colours. I have to tell you, once I realized what a difference eating breakfast made on my whole day, it changed the way I looked at eating. I realized that the times when I skipped eating first thing, when I gave in to the lure of the cosy duvet, were the times when I would have dark thoughts and feel confused. What I thought was confusion was really hunger, was really me ignoring my body's need for energy.

And then as soon as I ate, the sun would come out again, and my head and heart would clear.

BE YOUR OWN ENERGY EXPERT

Nowadays, breakfast is my favourite meal. It gives me the power and energy to get my trainers on so that I can do my morning workout, to help me plan my nutrition for the rest of the day. A few hours after breakfast, I'm already snacking to keep my energy up. I make sure to have a lunch that includes vegetables, grains, and chicken or fish. I have a snack in the afternoon, maybe a piece of chicken if I'm at home, or some rice and lentils that I've brought with me if I'm running around. And then my dinner takes into account what I've been doing that day, and how much energy I think I'll spend over the evening. In general I don't need as many carbs for energy at night as I do during the day, so I'll focus on vegetables and a smaller portion of protein for dinner.

How do you start your day? Do you eat a healthy breakfast, like muesli or eggs and vegetables? Do you carry healthy snacks, like nuts and fruit, to give yourself an energy boost between meals? Do you pack a healthy lunch if good options aren't available during the day? If you don't, you aren't giving yourself the energy you need to make it through the day. And energy is what makes the world run, whether you're a human, a laptop or a mobile phone.

I mean, really, you're already used to paying attention to energy. Don't believe me? Think about your mobile. Do you leave the house with a mobile that isn't charged and ready to go? Of course you don't. You always make sure it's powered up before you leave for the day. If you start running out of juice, and you haven't packed your recharger, you'll wind up frantically running around in search of a way to plug your phone into an energy source. In fact, the very idea of your phone powering down might freak you out enough that you want to look at it right now to make sure you have enough bars left to get your friends' texts and Instagram pics.

This is how you need to start thinking about your body's energy level. You must begin the day with a fully charged battery and develop an awareness of how your power level fluctuates throughout the day, and recharge when you need to. Eating whole foods and giving yourself all of your essential nutrients is like plugging in your iPhone: it charges up your whole life.

CHAPTER 5

EATING
THE SUN

T**HE STORY OF LIFE** on this planet is a story of energy: finding it, using it, caring for ourselves with it. So where does energy come from? Outer space?

Actually, yes. It comes from the sun.

The source of almost all the energy on our planet is the sun. We burn wood and coal for heat; the sun causes trees to grow, and coal is made of trees and plants that died millions of years ago. Fossil fuels like oil also come from organic life that was originally sustained by the sun. Wind power comes from the sun's rays changing the air pressure and shifting the air currents. And how about when you plug that phone in for a charge? Well, electricity is a refined version of energy from oil and coal and wind.

Without energy, you would be cold, you would be hungry and your phone would be dead. All of this energy, the energy that warms us, the energy that powers our phones, the energy that I eat, the energy that you eat – it all comes from the sun.

When I sit down in a restaurant and order a tomato and basil salad, I know that the growth of those tomatoes and that basil was powered by the sun. When a cow eats grass that grows in a field, it is eating the energy of the sun. In turn, when you eat a grass-fed burger with a side salad or a steak with grilled vegetables, you are eating the energy of the sun. That sun energy comes to us in the form of macronutrients – carbohydrates, proteins and fats – that give us our energy, our strength and our vitality. In short, nutrients from the sun give us life.

Since our energy sources come from the sun, it would be very convenient if we could just get our energy fix from lounging poolside for a few hours. Unfortunately, it doesn't work quite that way. Lucky for you, plants, algae and some bacteria called cyanobacteria have the amazing ability to turn that sun energy into their own life energy through a process called photosynthesis (photo = light, synthesis = putting together).

That plant energy is what makes cherries and beetroot so deliciously sweet, and is also what fuels all of your body's functions. When we eat plants and animals (who eat plants), we get sun energy in the form of macronutrients. There are three kinds of macronutrients: *carbohydrates*, *protein* and *fat*.

These macronutrients provide energy, and they also contain varying amounts of *micro*nutrients. There are two kinds of micronutrients: *vitamins* and *minerals*. Micronutrients give us all of the things you'd expect to find in a multivitamin – but when you get them from food, they're much more effective than a pill. Speaking of taking pills, let's not forget about water. While not considered a micronutrient or a macronutrient, water is considered a nutrient. Together the three macronutrients plus vitamins and minerals plus water add up to the six nutrients essential to our survival.

HOW SUNLIGHT BECOMES ENERGY

Have you ever wondered where carbs come from? Carbohydrates, the basic energy in plant food, are a combination of carbon dioxide, water and sunlight. As you may know, here on planet Earth, the air is a pleasant, life-supporting mix of nitrogen (four-fifths), oxygen (one-fifth), a bit of carbon dioxide and a few other gases. In order to make energy for themselves, plants use their leaves to absorb carbon dioxide (CO_2) from the air and their roots to absorb water (H_2O) from the soil. When those carbon dioxide and water molecules reach the surfaces of leaves and flowers, they are exposed to the light of the sun. That sun exposure causes a chemical reaction that breaks CO_2 and H_2O down to their most basic parts and reassembles them as carbohydrates, beginning with the simple sugar known as glucose.

THE BIG DEAL ABOUT MACRONUTRIENTS

You need a lot of macronutrients to stay alive. Every scrumptious bite of food that touches your lips is made up of either carbohydrates, protein or fat – or ideally, a combination of all three. These three macronutrients have different properties and different roles within the complex machine that is your body, but all of them deliver energy that your body can use as fuel.

Complex carbs, lean proteins and healthy fats that come from whole foods are all essential in the appropriate proportions, because together they contain the building blocks of your life. They nourish us at the cellular level.

Foods like rice and whole grains and vegetables provide carbohydrates, which your body turns into glucose, to give you energy. Fish and poultry and beans offer protein, which your body breaks down into amino acids to repair muscles. Healthy fats like nuts and olive oil give your body the essential fatty acids that it needs to absorb vitamins and minerals and keep you healthy.

Eating might sometimes feel like it's about satisfying a craving for something sweet or crunchy or salty, but the true purpose of food is to nourish your greedy little cells. Because all cells need fuel to survive. Even bacteria, the smallest of cells, need fuel – and human cells are about ten times the size of bacterial cells.

THE FUEL MILEAGE OF MACRONUTRIENTS

Macronutrients are to humans as petrol (or electricity) is to a car. When you fuel up your car, you have an idea of how many miles that petrol might provide if you are on a motorway or in the city. Carbs, protein and fat likewise come with energy rankings that help us determine how far they'll take us before we get hungry and need more nutrients to power us through.

The energy rankings we use for food are called calories. That's right, calories. Calories aren't a measure of how 'fattening' a food is, they're a measure of how much energy is packed into that food.

CARBOHYDRATES offer **4 CALORIES** per gram		

PROTEINS offer **4 CALORIES** per gram		

FATS offer **9 CALORIES** per gram		

As you can see, fats offer more calories per gram than proteins or carbs; for that reason, we say that they are energy dense. In other words, a little bit goes a long way. If you want to feel energized and whole, you want to eat carbs to get the fires burning, protein to keep you going steadily throughout the day, and fat to add flavour. You must also drink water to stay hydrated. By eating a variety of foods, you'll ensure that you're getting plenty of vitamins and minerals. By drinking plenty of water, you'll ensure that those vitamins and minerals make their way into your cells. Then your energy will be even all day long.

That's how I think about my food and my nutrition. I rarely think about calories, because calories are a funny thing. In the 1920s, the word was introduced to the public, and ever since, we have been obsessed. But calories are NOT an accurate measure of your nutrition. Calories give you energy, but if all you have is energy without the rest of your nutrition, you're not going to survive for long.

ALL CALORIES ARE NOT CREATED EQUAL

If macronutrients represent fuel, micronutrients represent QUALITY. They make the difference between the fuel that will power your life and the cheap stuff that will just clog your engine and slow you down. For example, grapes have calories *and* they have nutrition. Grape fizzy drink *only* has calories and no nutrition. Same goes for tomatoes versus tomato ketchup, apples versus Applejacks, or the black bean tacos you make at home with red cabbage salad and fresh salsa versus the drive-thru tube you pick up on the way home from work.

Natural foods that grow in the ground contain vitamins and minerals. Processed foods sometimes have vitamins and minerals *added*, but the very act of processing whole foods strips out their nutrients and fibre. When nutritionists talk about the difference between whole and processed foods, they use the terms *nutrient-dense calories* and *empty calories*. The more nutrients a calorie offers, the more nutrient dense it is. Empty calories are, well, EMPTY CALORIES. Processed junk doesn't offer quality nutrients . . . so all you get are the calories.

What does this mean for you?

It means that you can down a day's worth of calories – sometimes in one sitting – without getting *any of the nutrients your body needs*. While your belly feels full (maybe too full), your cells haven't obtained what they need to function, to keep you feeling good, to power your health.

If you respond to the feeling of hunger by eating high-calorie, nutrient-poor foods like processed foods, fast foods and overly sweetened desserts, you're not 'indulging' – you're denying yourself the nutrition you really need to thrive. But when you eat cherries instead of cherry pie, grapes instead of grape drink, carrots instead of carrot cake and homemade tacos instead of drive-thru burritos – in other words, nutrient-dense foods – every bite you eat powers up your engine.

SAY NO TO LOW

Over the past few decades of diet and weight-loss crazes, each of the macronutrients has been subjected to smear campaigns. First, in the 1980s, people turned against fat, claiming that it was causing disease and weight gain. Do you remember all of those fat-free snacks on the market? Biscuits, ice cream, crackers, even fat-free cheese, for goodness' sake. Well, nothing's for free, baby.

When they took out the fat, they replaced it with sugar, which just made people fat. Not to mention that fats (whole-food, healthy fats) are actually *good for you*. Then carbs took it on the chin, and everyone went on a low-carb diet (which, ironically, tended to be a high-fat diet), forgetting that complex carbs and whole grains like brown rice or quinoa are not the same carbs you'll find in chips or pizza. Most recently there's been a trend towards cutting out all animal protein, which again, is healthy in moderate amounts (and a nice piece of wild salmon is a very different source of protein than my old drive-thru standbys).

These trends have been confusing, and they've also been dangerous for our bodies and minds. By the age of thirty, most of us have absorbed enough misinformation that we're not sure what to believe anymore. Let's face it: if diet trends were effective, we'd all be eating fat-free, low-carb biscuits in our bikinis instead of throwing money away on one diet programme after another. It's time to relearn our basic biology.

THE MICRONUTRIENTS:
SMALL AMOUNTS WITH BIG RESULTS

You don't need micronutrients in the same quantity as macronutrients, but they are still essential to your health. If you eat a variety of fruits and vegetables daily, you're probably getting plenty of vitamins and minerals. If you aren't eating fresh produce on a regular basis, you need to get those salads going. And make sure that they're loaded up with more than iceberg lettuce!

As we'll discuss in a few chapters, micronutrients are our indispensable allies, and the more you eat a vivid, colourful array of fresh foods, the more vital and alive you will feel. If your meals are lacking in fresh fruit and vegetables, if you become deficient in one or more vitamins or minerals, any number of conditions can arise, from depression to muscle degeneration to hair loss. Sound serious? It is. But you can sidestep those outcomes by enlivening your plate with bright green greens, bold reds and sunny golden hues.

Our growth as children relied on our intake of micronutrients, and so does our health as adults. The health of your bones, muscles, vision, brain function, immune system – all of these are dependent on the micronutrients, vitamins and minerals that you consume with every meal.

AND THE MOST IMPORTANT NUTRIENT OF ALL …
WATER

Water has no calories, but it may be our most important nutrient of all because it is an essential part of all of the chemical reactions that give you life. It doesn't give you fuel. But those two hydrogen molecules paired with one oxygen molecule play some amazing roles in your health. Water helps regulate body temperature by acting as a coolant for your internal system. It transports nutrients to your cells. It whisks away waste products. Without water, you couldn't turn those carbs, proteins and fats into the usable energy that powers every breath.

VEG-FRIENDLY NUTRITION

Vegetarians get their energy from plants only. There are many reasons why people might opt for a veg diet, whether they love cows too much to eat them or just really, really love the taste of vegetables. They may believe it's better for their health, better for their souls, better for the planet or better for their wallets.

Within the meatless kingdom, there is more than one eating style.

- Some people eat dairy and eggs along with their fruit and vegetables (lacto-ovo-vegetarians).
- Some eat dairy, not eggs (lacto-vegetarians).
- Some eat eggs, but no dairy (ovo-vegetarians).
- Some eat plants, abstaining from any product that has been derived from an animal, including eggs, dairy and honey (vegans).

I definitely advise befriending a vegetarian. Seriously, some vegetarians have developed a real knack for turning garden-fresh produce into delectable, satisfying, I'd-have-thirds-if-I-wasn't-already-so-full kind of meals. As a chicken- and meat-eater who loves my vegetables, I find that my vegetarian friends often introduce me to new varieties of veggies and show me amazing ways to cook and prepare them.

In the end, whether you are dabbling in vegetarianism or you've been eating that way for decades, you still face the same challenges as everybody else: making sure you consume a variety of whole foods to provide your body with the nutrients it needs.

CHAPTER 6

COMPLEX CARBS ARE ENERGY

———

WHEN I LOOK AT a bowl of porridge, a side of roasted sweet potatoes or a gorgeous yellow ear of corn on the cob, I think of energy. When I look at a bowl of brown rice or quinoa (an ingredient that often puzzles people in terms of preparation and pronunciation – the answers are, it cooks in twelve minutes and is pronounced 'KEEN-wah'), I see the kind of foods that make my taste buds very happy – and I see my ability to work twelve-hour days on set making a movie or to have the energy to go surfing, walk around the city, hang out with friends or do anything else I feel like doing, including writing this book. Everything we do requires physical energy and brainpower.

I love being active and being able to be mentally present and aware. And therefore, I love carbs. I love them! LOVE THEM!! They are what give me the energy I need to do all the things that I love to do. Choosing the *right* kind of carbohydrates allows me to get the most out of my day. The right carbs are whole-food sources: foods like hearty grains that are still the way nature made them, not refined into a pizza crust or a pretzel twist, whole fruits that haven't been turned into juice and aren't swimming in sugar, and vegetables that are eaten fresh or cooked with some olive oil or another healthy fat.

YOUR BRAIN ON CARBS

Your brain, nervous system and red blood cells rely on carbohydrates as their major fuel source. Your muscles use carbs for fuel when you do anything that requires a little burn. Whether you're playing tennis, riding a bike or dancing up a storm; whether you're writing an exam, writing a screenplay or loading data into a computer, you need carbs in order to keep going.

In fact, dietitians and nutritionists generally recommend that carbs provide about half (45–65 per cent) of our total daily energy (or calories).

So why are people so afraid of carbs?

Well, somewhere along the line (probably due to a diet craze), people started confusing complex carbs – which are found in whole grains, vegetables, fruits and pulses (aka beans) – with simple carbs, like those found in refined grain products and sugary junk foods. Complex carbs support your energetic, healthy life, while simple carbs are a source of empty calories. Additionally, simple carbs are sneaky, because when you eat something that's been refined and doesn't have any fibre, you can eat and eat and never feel full, which means you can gain weight without even knowing that you're overeating . . . which double sucks. I know that if I eat a big bowl of white, refined pasta with only tomato sauce on it, I'll get tired immediately after eating it, and I'll be hungry again shortly after that. But if I eat a bowl of brown rice pasta with some sautéed broccoli and courgettes and grilled chicken, I'll be pushing through the next few hours with energy as well as a satisfied belly. And if I want to satisfy my energy needs and hunger even more, I would make the meal with whole-grain brown rice or quinoa. Because whole grains aren't processed at all, and as we'll discuss, whole grains give you a longer-lasting energy source than refined carbohydrates.

THE WHOLE TRUTH ABOUT WHOLE GRAINS

A whole grain is a seed from a plant: the whole seed, with nothing added and nothing taken away. A seed has a few parts: there is the endosperm, the bran and the germ. The endosperm is very starchy and doesn't really offer much nutrition. The bran, the layer on the outside of the seed or grain, contains fibre. The germ is where the nutrients live, like iron and the B vitamins, which include niacin.

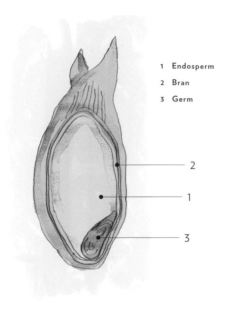

1 Endosperm

2 Bran

3 Germ

2

1

3

Whole grains can be eaten as whole seeds, like brown rice, or they may be *cracked*, like cracked wheat, or bulgur. They can also be ground into flour that's used to make everything from pasta to cereal to pizza dough. If you take whole-wheat berries and grind them into flour, you end up with whole-wheat flour that is made of all the parts of the grain: starchy endosperm, fibre-rich bran and nutrient-rich germ.

You know the plain flour that most people use when they make chocolate chip cookies? Well, that's refined wheat flour. Even though it began its life as a plant in a field, even though it started as a seed with an endosperm, bran and germ, by the time it ends up in a biscuit, it's been processed to remove the bran and the germ, leaving only the endosperm. Got that? The fibre and the nutrition have been stripped out, leaving only the starchy endosperm. And even that is typically bleached so that it's pretty and uniformly snow-white. Then some nutrition is injected back in, like thiamine, folic acid, niacin and iron. How strange is that? We remove the nutrition that came packaged in that perfect little grain. Take away the good stuff, then try to replace some of what was taken away. Crazy.

That's the difference between REFINED grains and WHOLE grains.

Think about rice. You know the soft white rice that comes with your Chinese takeaway? That rice has been stripped of its bran and germ, leaving only the starchy, not-very-nutritious endosperm for you to eat with your broccoli in garlic sauce. If you order brown rice, you're getting fibre and nutrition along with your energy. And the energy you get is a slow, steady energy instead of a quick rush. Why? Because your body can process simple carbs easily – *too* easily. When you eat simple carbs, which are single molecules, they are so usable to your body that all of the energy is absorbed at once – giving you a short-lived rush. Complex carbs are made up of a string of connected carb molecules. Your body works hard to detach the molecules for use, which takes time, giving you a longer-lasting energy source.

Simple carbs are made of one or two sugar molecules.
Complex carbs are linked chains of three or more.

SIMPLE CARB

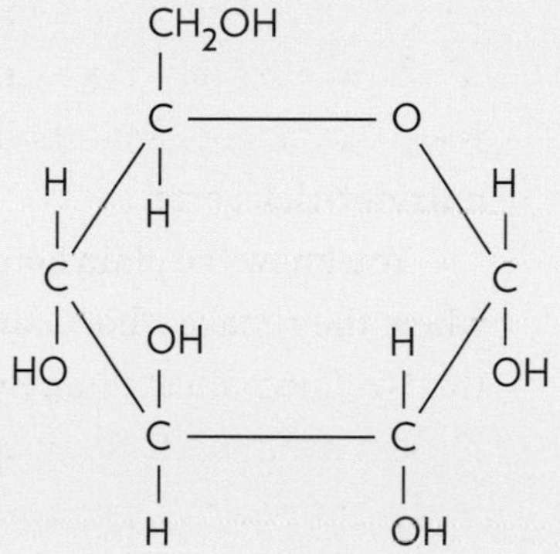

CH_2OH

C ——— O

H

H

C

C

HO

OH

H

OH

C ——— C

H

OH

COMPLEX CARB

OH

OH

HO

O

O

OH

OH

O

OH

OH

HO HO

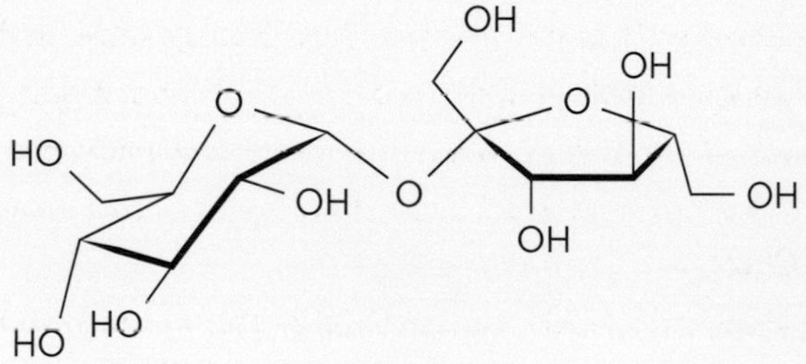

With that in mind, doesn't it make more sense to just eat and enjoy whole grains rather than eating 'fortified' processed grains? You get the fibre. You get the nutrition. You get a variety of different foods to eat and enjoy, and your body gets long-lasting energy. With whole grains, your plate will be full of colour and you'll be full of pep. Brown rice and red rice and black rice. Wild rice. Wheat berries. Millet. Oats. YUM!

There are so many delicious complex-carb choices out there. Here's a short list of the ones that you can always find in my kitchen and on my plate!

FRUIT	VEGETABLES	GRAINS	BEANS	PASTA
grapefruit	kale	quinoa	chickpeas	brown rice pasta
tomatoes	spinach	brown rice	black beans	couscous
apples	sweet potatoes	steel-cut oats	lentils	quinoa pasta
			pinto beans	

WHAT IS FIBRE, ANYWAY?

People talk a lot about fibre when it comes to health and weight loss – so let's look at what fibre actually is and what it actually does.

Fibre is a kind of complex carbohydrate that humans can't digest, found in plants like fruits, vegetables and grains. Even though our bodies can't break it down, it's important for us to eat plenty of fibre-rich food, because that very stuff that we don't digest is the stuff that exercises our digestive tract and helps to eliminate waste from the body. Fibre also reduces our risk of diseases like diabetes, heart disease and colon cancer, helps promote a healthy weight and keeps your cholesterol levels in check.

The best sources of fibre don't come powdered in a jar or injected into a cracker or biscuit (or a beverage!). They are simply whole, delicious foods. When you eat a diet that's mostly made up of the kinds of nutritious foods we've been talking about, you can't help but get your fabulous fibre.

How much fibre is enough? It's estimated that back when we hunted and gathered our food, we ate 100 grams of fibre every day. These days, the recommended allowance for young and adult women is one-quarter of that – 25 grams (26 grams for teens). The consumption of fast and processed foods is one reason why we're eating less fibre – so in order to meet your daily requirement, it is important to eat your food the way nature packaged it for

UNHAPPY MEALS

The texture of frozen food is better preserved if there isn't much fibre in the food. Fast-food manufacturers make sure their frozen fries, patties, buns, everything – are low in fibre so that their unhappy meals taste exactly the same, every time.

you. And nature also gave us a wonderful tool – our teeth – to help kick-start the beginning stages of digestion. As your teeth churn and mash your food, that crushing action, along with the amylase enzymes in your saliva, begins to break down fibre so that as the food travels down the rest of your digestive tract, your body has a better chance of extracting the nutrients from it. So your mum was right – you should always chew your food well!

If you ate porridge for breakfast today, snacked on an apple midmorning, and had a bowl of black bean soup with a side salad for lunch, you'd already be more than halfway to your fibre goal. And that's good news for your energy levels, too, because fibre helps regulate your blood sugar, keeping you on an even keel instead of sending you on the high/low sugar roller coaster of processed carbohydrates.

ENZYMES are proteins that help your body build molecules and break down molecules. Enzymes are used for processes like digestion – and, according to *Gulp* by Mary Roach, can also be found in laundry detergents, because they 'digest' food stains to remove them!

THE TWO TYPES OF FIBRE

INSOLUBLE FIBRE

- **Also known as:** cellulose, hemicellulose.
- **Where it's found:** whole grains, like oats and barley and wheat; seeds and nuts; vegetables, like courgettes and celery; fruit, like grapes and sultanas.
- **What it does:** Because insoluble fibre cannot be digested by the body, it helps move food and waste through your digestive system (like a loofah for your insides). I remember hearing once that broccoli was nature's broom. This always makes me chuckle, because it's true! As your body processes food down into nutrients, it can use the fibrous broccoli tops that aren't digested to sweep out the rest of the waste in your colon. Kind of awesome!

SOLUBLE FIBRE

- **Also known as:** pectin, 'gums', mucilage, psyllium and lignins.
- **Where it's found:** grains, like oats; pulses, like lentils and beans; fruits, like apples and oranges; vegetables, like cucumbers and carrots.
- **What it does:** Soluble fibre is absorbed in your body and slows digestion for maximum nutrient absorption. If and when you get bloated or gassy after eating a high-fibre meal, it's because the bacteria in your system are turning the soluble fibre into gas.

TO GLUTEN OR NOT TO GLUTEN

That basket of warm, fluffy bread on the table has become a hot dinner-party topic these days. Many people are passing on it not because of a fear of carbs, but because of concerns about gluten. So what is gluten, exactly?

Gluten is Latin for 'glue', and it's a protein found in wheat and other grains that gives bread its chewy texture. According to the Whole Grains Council, all wheat contains gluten, including spelt, kamut, faro, durum, bulgur and semolina, as well as barley, rye and triticale. Grains that don't contain gluten include amaranth, buckwheat, corn, millet, oats, quinoa and rice.

Some people avoid gluten because of its link to chronic inflammation (see page 52), while others skip it due to gluten sensitivities or an intolerance called coeliac disease. A gluten allergy can cause anything from indigestion to rashes, depression and joint pain. The National Foundation for Celiac Awareness says that while 3 million Americans have the disease, only 5 per cent are diagnosed. If you think you may have a sensitivity, talk to your GP.

FUN WITH CARBOHYDRATES

I'm always looking for new ways to build flavour and have fun with my carbs. For instance, I prefer savoury over sweet, so in the mornings when I make porridge, instead of eating it with something sweet, I make a yummy concoction of sautéed courgettes with spring greens, caramelized shallots and egg whites. I top it all off with ponzu sauce (a Japanese sauce made from rice vinegar and citrus) or some lemon juice. It's a breakfast that is delicious and savoury and all the things that I love to taste. I created this dish so that I could eat porridge in the morning, because it is an awesome source of complex carbohydrates.

Same thing goes with that bowl of pasta I want to eat for lunch. I'll eat it, oh I WILL eat it. But instead of eating white-flour pasta, I'll make a whole-wheat pasta or my favourite quinoa pasta, and I'll sauté some spinach with fresh tomato and garlic or shallots to top it off and finish it with a little Parmesan and lemon. I'm still getting the pasta I'm craving as well as getting the nutrients of the spinach, tomato, garlic and lemon juice.

I love that along with the enjoyment I get from a delicious bowl of pasta, the carbs I am eating are also giving my body and brain the nutrients they need to fire on all cylinders. You can create any experience you want with the food that you eat.

CHAPTER 7

SUGAR IS NOT A NUTRIENT

D O YOU LOVE SWEET foods? If you're like most people, the answer is yes. And it turns out, there's a biological reason for that: sweet foods aren't poisonous.

Old-school humans loved sweet foods because our ancestors knew that if a food was sweet, it was safe to eat. Sweetness was a sign that a plant was edible (most plants that are poisonous to humans taste bitter). Sweetness is also an indication that the plant is high in glucose, which meant that it would offer us lots of energy. This is a natural response, and we know this because babies love sweet foods. As parents try to diversify a baby's diet, they generally have to offer their kids a new food at least ten times before the child will eat it. But if the food is sweet, he or she will go for it on the first try (and will probably ask for MORE!).

So yes, humans love sweets, and that love is natural and pure – when you're talking apricots and cherries and watermelons and muskmelons. The sugar contained in fruit, fructose, is a healthy sugar when you're eating the WHOLE fruit (usually including the skin) and getting all of the fibre and vitamins and minerals of the fruit. But when sugar is extracted from sweet plants and added to another food (like bread or cereal), you're not getting any nutritional benefits. That sugar is nothing but an ADDED SUGAR, and it offers nothing but empty calories.

Dentists hate sugar because it rots your teeth, doctors warn against sugar because it is linked to obesity, and infant school teachers fear sugar because it makes children bounce off the walls. Your individual relationship with sugar is one that you should identify and reflect upon. Personally? I think that you should consider breaking up with sugar. Immediately.

But before I continue, I've got a confession. It's a big thing to admit in public, actually, because most people don't feel the same way.

Here it is in print, where I can't take it back: I don't like sugar.

I really don't. Sugary things do *not* make me go back for more. Salty, fatty, greasy, yes. Sugary, no.

There, I've said it. It feels good to get that off my chest. All my life, people have looked at me incredulously when I gently decline their shared treats. "Not even this amazing candy?" Nope, sorry, not into candy. "How about this delicious raspberry jam?" No, thank you, unless you spread it on a salty buttermilk scone and douse it with clotted cream. Or unless you can turn it back into raspberries, which I love by the handful. "Not even *ice cream?!*" they'll ask. OK, you got me there – I do love ice cream. But only if it has a savoury element, like salted caramel ice cream.

So there you have it: I do not like sugary foods. And the more I learn about the dangers of sugar, the more grateful I am that I'm not drawn to it.

The sugar habit is a bad one – I've seen it hurt friends, contributing to anything from food addiction to diabetes to that extra stone around the middle they're always talking about wanting to get rid of. The more I learn about all of these things, the more I realize that my natural aversion to sugar is probably one of the pillars of my health.

SUGAR, SUGAR EVERYWHERE

There are many different kinds of sugar: the sugar in milk, the sugar in fruits and vegetables, the sugar that sits around in little packs in the coffee shop. These sugars are not all created equal!! Herewith a list of some common sugars, both naturally occurring and processed (which become added sugars in packaged foods):

GLUCOSE: Glucose is present in nearly everything you put into your mouth, from fruits and vegetables to biscuits, cakes, and candies. It is in breakfast cereal. It is in cheese. We've already talked a lot about glucose, and we're going to keep talking about it, because glucose is the most abundant sugar found in foods, and it is used as a source of fuel by all organisms, including us. When you eat complex carbohydrates, your body turns those whole foods into the glucose that you need to live.

FRUCTOSE: Fructose is fruit sugar. When you eat fibre-rich fruit that also contains fructose, your body absorbs the sugar for energy, and the fibre keeps the sugar from overloading your system because it slows down your digestion. That's why I eat fruit. I eat apples all the time. Fruit is nature's energy-delivery system – offering delicious sweetness in a formula that sustains us.

So there you have it: I do not like sugary foods. And the more I learn about the dangers of sugar, the more grateful I am that I'm not drawn to it.

SUCROSE: Sucrose is the little pot of fine, white sugar that most people stir into coffee. It's also found in brown sugar, honey, maple syrup and molasses. Sucrose is actually a combination of glucose and fructose. When you add sucrose to food, you are adding sweetness that either enhances or masks its flavour. But you are also adding calories and taxing your digestive system. Studies have also shown that when you ingest sucrose, the sugars bypass the hormones that tell you that you're full, which means you can overeat without realizing it.

THE RELATIONSHIP BETWEEN INSULIN AND SUGAR

Insulin is a hormone that helps deliver glucose to your cells. As sugar enters your bloodstream from the food you eat, your pancreas secretes insulin, which regulates your blood sugar by transporting the glucose out of your blood and into your cells. When you eat a lot of sugar, your pancreas is forced to go into overdrive, producing high levels of insulin to compensate. If you regularly eat a lot of sugar, this elevated level of insulin, over time, can lead to a condition known as *insulin resistance*. When you become insulin resistant, your cells are less responsive to the presence of insulin; as a result, they need more insulin to absorb glucose from your blood. So your pancreas pumps out even *more* insulin, again and again. Insulin resistance has been linked to the development of heart disease and is a precursor of type 2 diabetes.

Over the past two hundred years, people in the developed world have fallen hard for sugar. And the results are killing us. Eating added sugars makes you gain weight. It triggers your body to store fat in your belly. It bypasses your natural hormonal 'I'm full' system and prompts you to overeat. It tricks your brain into thinking you are hungry for MORE. And it leads to obesity, heart disease and diabetes.

The numbers are staggering! To take the United States for example:

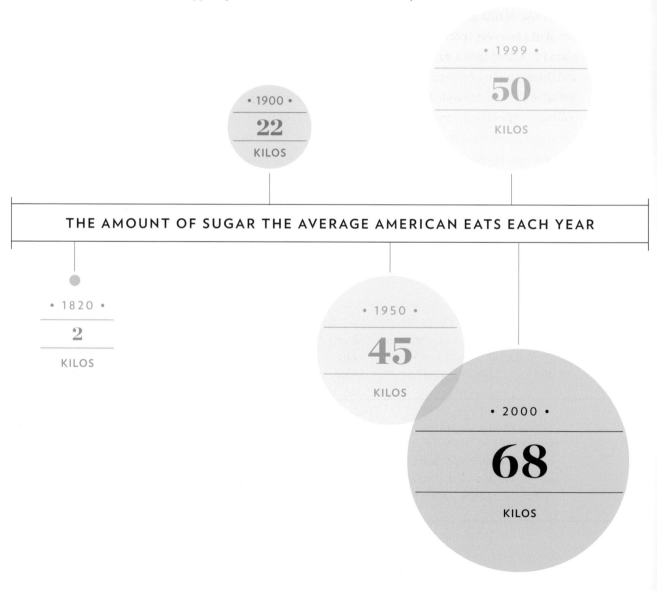

• 1999 •

50

KILOS

• 1900 •

22

KILOS

THE AMOUNT OF SUGAR THE AVERAGE AMERICAN EATS EACH YEAR

• 1820 •

2

KILOS

• 1950 •

45

KILOS

• 2000 •

68

KILOS

HOW SUGAR BECOMES SUGAR

Pure white, granulated, easy-to-pour white sugar comes from a plant: either sugar cane or sugar beets. Sugar cane in its natural form – tough, thick, tall stalks that you have to work very hard to get at – can be delicious. Once you get past the cane's tough exterior, you'll find a chewy fibre that you can't bite off, with a light, watery, sweet juice that you can suck out of the fibres as you gnaw on the stick. Sugar beets look like white roots, and they grow beneath the ground.

Those plants are a far cry from the bags of table sugar that are the end result of a few very intense rounds of processing. The juice extracted from the plants is boiled into a syrup, which is evaporated until it turns into crystals and then spun in a huge centrifuge, where the wet parts are extracted, leaving lighter-coloured crystals behind. Then the process is repeated twice more: boiling, evaporating, crystallizing, spinning. Molasses – a very dark, thick, sweet syrup used in cooking and baking – is the stuff that emerges when boiled, crystallized cane juice or sugar beet juice is spun in a centrifuge. 'Raw' or 'turbinado' sugar is sugar that has been processed one less time than refined white sugar, so some of the molasses colour remains. It isn't really raw, though, as it's been boiled and crystallized and spun at least twice!

HOW SWEET IT ISN'T

We've all seen the brightly coloured packs of not-sugar near actual sugar at the coffee shop. If you're getting rid of your sugar habit, don't add another habit by swapping sugar for these nonnutritive (meaning, *no nutrition*) sweeteners. Just don't go there. Nonnutritive sweeteners, artificial sweeteners, and low-calorie sweeteners are often added to diet soft drinks and foods like light yoghurt, sugar-free pudding and sugar-free candy to increase the sweetness without adding calories.

I say, don't eat them. Training your taste buds to appreciate the delicate sweet flavours of natural fruits is the way to keep your body healthy and enjoy your food. And besides, those sweeteners admit what they really are: ARTIFICIAL sweeteners. They're faking it. They contain a bunch of chemicals that may have originated in nature long ago but are no longer being used in the way nature intended. If you really need to have something sweet and fruit just won't satisfy, I'd rather see you use plain refined sugar than those man-made fakes.

THE DANGERS OF INFLAMMATION

Perhaps you've heard people talking about inflammation and how dangerous it is for your long-term health. Well, there are two kinds of inflammation – one that is helpful to your body, and one that is damaging and dangerous. Acute inflammation pops up when you cut yourself or get a sore throat. The resulting swelling is an example of inflammation, the lifesaving response your immune system triggers to protect you. Your immune system is like your body's security force. When it senses the presence of an intruder, it sends an army of white blood cells to a specific area to protect you from harm. That protective inflammatory response ensures that your little paper cut doesn't become an infected wound.

The other kind of inflammation is chronic inflammation, and some doctors believe that it creates an environment in which diseases – from the obesity, diabetes, and heart-disease gang we keep talking about to illnesses like depression and cancer – can thrive. Chronic inflammation is linked to eating processed foods, added sugars and not getting enough exercise.

Regular physical activity of moderate intensity can strengthen your immune system, protecting you against colds and other infections. Intense physical activity can sometimes stimulate an immune response, which can contribute to chronic inflammation (we'll talk more about this in Chapter 20). How can you protect yourself against chronic inflammation?

- **Get off the couch.** Being sedentary – especially for women – has been found to increase biomarkers (molecules whose presence indicates the likelihood of developing various diseases) for inflammation.
- **Increase your consumption of fruits and veggies,** especially those rich in vitamin C and beta-carotene. These antioxidant-rich nutrients may help your body minimize the stress response.
- **Increase your consumption of omega-3 fatty acids.** Include fish in your diet a couple of times a week. Add some walnuts to your porridge at breakfast. Put a few slices of avocado on a salad at lunchtime.
- **Get enough sleep.** Lack of sleep may be associated with more inflammation. Aim for seven to nine hours per night.
- **Avoid excess body weight, especially around your abdomen.** Tummy fat is linked more closely with excess inflammation than other kinds of body fat.
- **Revise your workouts during stressful times.** High-intensity workouts of long duration may be counterproductive and result in overtraining, characterized by lots of inflammation. Moderate-intensity workouts, such as riding your bike for an hour, may be better and actually help control inflammation.
- **Maintain a positive outlook!** Reducing stress levels is also an important part of preventing inflammation.

THE BIG DEAL ABOUT HIGH-FRUCTOSE CORN SYRUP

So there's processed sugar, and then there's high-fructose corn syrup (HFCS). Which comes from . . . corn. Which, as you probably know, is sweet, but it's not *that* sweet. So how do they turn it into the sticky stuff that manufacturers have been dumping into our foods for the past few decades?

Well, let's start in the '70s, when manufacturers wanted to spend less money on the ingredients for their processed foods and drinks. So much corn is grown in the United States that it made financial sense to use it as a source of sugar. The only hitch was that corn syrup wasn't quite as sweet as sugar . . . so manufacturers created *high-fructose* corn syrup. It's corn syrup on steroids.

The use and production of high-fructose corn syrup has increased dramatically in the past three decades, and gets lots of attention in the media because the rise of HFCS coincided with the rapid increase in obesity in the United States. The important thing to remember is that *all* added sugars, whether table sugar (sucrose), high-fructose corn syrup, maple syrup or honey, can cause damage to your body over the long term.

READ THE LABELS

To stay ahead of (and away from) added sugars, always check out the nutrition facts panel of a food label. You'll see 'total sugars' listed, but that's not the best way to tell if a food contains added sugars. Here's why: because 'total sugars' includes *all of* the sugars in the food, naturally occurring and added, and we're only trying to stay away from *added* sugars. If you're looking at a label for a type of food that doesn't contain naturally occurring sugars from fruit or milk, then the sugars noted would be added sugars. But if you're looking at the nutrition label of a fruit or dairy product, like applesauce or yoghurt, the number you see is the total sum of both natural and added sugars.

Be aware that food manufacturers often try to conceal just how much sugar you're eating by using many different kinds of sugars so that it's less obvious that the main ingredient is actually SUGAR.

Here are a few variations to look out for:

agave nectar	evaporated cane juice	lactose
brown sugar	fructose	maltose
cane crystals	fruit juice concentrates	malt syrup
cane sugar	glucose	molasses
corn sweetener	golden syrup	raw sugar
corn syrup	high-fructose corn syrup	sucrose
crystalline fructose	honey	syrup
dextrose	invert sugar	

HOW ADDED SUGAR ADDS UP

It's typically a good idea to choose healthful-sounding foods like salads, fruit and yoghurt. But you've also got to make sure that these foods don't just *sound* healthy. Added sugars can turn something that is wholesome into something that isn't. For instance, plain yoghurt naturally has seventeen grams of sugar, which is okay, because that sugar comes from lactose, which is a naturally occurring sugar. But when you get the fruit-added yoghurt, which almost always means sugar added, you can get up to forty-seven grams of sugar, which is thirty grams of ADDED sugar. If you have plain yoghurt with a handful of blueberries, you get your yoghurt, your fruit, your sweet treat with no added sugar – and everybody is happy. Here are a few more examples of how added sugar adds up.

TYPE OF FOOD	WHOLE FOOD VERSION	WITH ADDED SUGAR
Salad dressing (1 tbsp)	Oil and vinegar 0.5 g	Thousand Island 2.5 g
Yoghurt (250g)	Plain 17 g	Fruit on the bottom 47 g
Liquid refreshment (250ml)	Plain water 0 g	Gatorade 13 g
Instant porridge (1 serving)	Plain 0.5 g	Raisin & Spice 15 g
Peanut butter (2 tbsp)	Natural 2 g	Skippy super chunk 4 g

HOW TO AVOID ADDED SUGAR

I believe that we can learn to skip added sugar unless we are choosing to indulge in a bite of a particularly decadent and delicious treat. Here are some strategies to help combat your sugar habit:

- **Cut back on the sugar *you add*.** Stop pouring sugar onto the foods you eat and prepare, like cereal, porridge, coffee and tea.
- **Do not drink sugar-sweetened beverages.** Just don't do it!! A 600-ml sports drink or sweetened flavoured water contains around 30g (that's seven teaspoons of added sugar).
- **Spice it up.** Learn to use spices like cinnamon, nutmeg and cardamom, and flavourings like vanilla to add sweetness and appeal, so you don't need the sugar. An apple sliced and sprinkled with cardamom is a surprisingly rich-tasting snack!
- **Choose fruit over sweets.** Have you ever noticed that most sweets come in fruit colours and fruit flavours? Sweets are basically fake fruit that wishes it was fruit and is doing its best to smell and taste like it. I mean, seriously: an orange in season is just as sweet, and has even more orange flavour than that lollipop, because it's *actually an orange*.
- **Watch out for imposters.** Be on the lookout for these common added sugar culprits. Just because they don't taste sweet doesn't mean they aren't spiked with corn syrup or other added sugars. Foods like:

ketchup	cereal	barbecue sauce
pretzels	energy bars	yoghurt
conventional nut butters	sports drinks	salad dressing
	instant porridge	granola
cereal bars	spaghetti sauce	

I love to eat fruit, but I will *never* put a spoonful of white sugar or brown sugar or even agave in my green tea. Yes, I like the taste of tea as is. But more important, those familiar-looking spoonfuls aren't as innocuous as they look.

Without the benefit of fibre to slow things down, sugar enters the bloodstream in a rush. Your body responds by sending a bunch of insulin to help that glucose get absorbed by your cells where it can be used for energy. You experience this cellular process as the heady feeling you get after you eat a pile of sweets – followed by the energy slump that leaves you longing for a nap, along with the mood swing that makes you question your own existence, not to mention the stomachache that fills you with regret.

So how much sugar is too much sugar? Different eating plans offer varying guidelines. In the UK, the NHS currently recommends that women consume no more than 50g of added sugar a day. Meanwhile, the American Heart Association suggests that women shouldn't eat more than 25g (six teaspoons) of added sugar each day. Let's put that into perspective: one 600-ml bottle of a fizzy soft drink contains nearly 85g (twenty teaspoons) of sugar.

My personal guideline? I'll share a dessert with friends every once in a while. But on a daily basis, fruit is my sweet treat.

CHAPTER 8

PROTEIN IS STRENGTH

W**HEN I THINK OF** protein, I think of barbecues, of Cuban feasts, and my favourite snack of quinoa, lentils and brown rice. I love to have a yummy roasted chicken for dinner – or for breakfast! For lunch, maybe a light and delicious piece of grilled fish . . . I also adore having a big pot of black beans going on the stove, perfect for making tacos or enjoying with a side of brown rice. In fact, I love all kinds of beans, and of course – as anyone who has ever set foot in my house knows – lentils of every colour. And as anyone who has ever come over for breakfast knows, they'll be offered one preparation or another of eggs. Scrambled, fried, sunny-side up, in a frittata . . . any way you can think of, eggs *will be* served!

The word **protein** means 'of prime importance' – and protein is certainly important to our health. Proteins are made up of amino acids. Amino acids and proteins are so vital to our health that they are regularly referred to as 'the building blocks of life'.

THE AMINO ACID TRIP

When you eat protein, your body breaks it down into its smallest components, amino acids. Amino acids are molecules that are used to build and repair all of the cells in your body, including your DNA. Scientists disagree on the exact number of amino acids that exist, but it's thought that there are about twenty amino acids that combine to make proteins; as an adult woman, your body can

PROTEIN FOR PLANTS

Plants need protein too. Plant protein requires nitrogen, which plants must absorb from the soil. (Even though four-fifths of the air on this planet is nitrogen, plants can't absorb nitrogen from the air.) And protein is very important for plants, in part because it helps give their stems structural support so they can lean towards the sunlight and can get their glucose fix!

formulate nearly two-thirds of them. Two-thirds is good, but unfortunately, it's just not good enough. There are eight amino acids that we *must* get from nutrition, and these are the **essential** amino acids. Every single day, each of these essential amino acids is required for your body to maintain your health.

As we've discussed, when you eat carbs and fats, your body can store away some of the excess for later use (whether you like it or not). But protein is different. Amino acids cannot be stored in the body. So the best way to eat protein isn't to sit down to a giant caveman-style slab of meat at the end of the day, but to eat small portions of protein steadily, throughout the day. That way, the amino acids are always available for your body to use.

HOW MUCH IS ENOUGH?

When you hear people talking about nutrition, everyone's always talking about getting 'enough' protein. What does 'enough' look like?

Protein should make up roughly 35 per cent of your total daily calories, which means that one-third of your breakfast bowl, your lunchbox, and your dinner plate should be a healthy protein. Children, teens and pregnant women need even more, because protein supports growth and development. For the rest of us, eating enough protein will ensure that we can build and maintain our muscles and our bones, as well as the antibodies that keep us healthy, the hormones that affect our mood and the enzymes that make digestion possible. Without protein, you wouldn't be able to go on a dinner date, fall in love, throw your arms around someone's neck and feel your heart race as you kissed.

The amount of protein each of us needs is a little bit different, because our protein requirements are based on our body weight and our physical activity. After all, it makes sense that it wouldn't be a one-size-fits-all equation, because the more or less you weigh, the more or fewer cells you have. And if you're super physically active – let's say you're training for a marathon, for example – your body is constantly building and repairing muscle, so you're going to need more protein than someone who runs five kilometres a few times a week.

Basically, the more active you are, the more strength training you do, and the more your body weighs, the more protein you need to support your form and let your body heal and rebuild. The chart opposite will tell you how much protein you want to be eating, based on your activity and on your body weight.

THE SKINNY ON SUPPLEMENTS

Protein supplements may be trendy with athletes who are looking to build and repair muscle quickly, but most healthy individuals – even vegetarians! – don't need to supplement with protein. Kathleen Woolf, a professor at NYU's Steinhardt School of Nutrition, advises women to eat protein steadily throughout the day. "If you're eating a variety of protein-rich foods, it is not likely that you'll need an additional supplement," she says.

IF YOU ARE LIGHTLY ACTIVE

If your fitness habit includes two half-hour runs every week, plus an hour of yoga and half an hour of weights, for a total of two and a half hours over the course of the week, you can consider yourself to be lightly active. Whether you do five thirty-minute sessions or three fifty-minute sessions, lightly active means that you work out for around two and a half hours over the course of a week.

YOUR WEIGHT	YOUR DAILY PROTEIN REQUIREMENT
7 stone	36 g
8 stone	42 g
9 stone	47 g
10 stone	53 g
11½ stone	58 g
12½ stone	64 g

IF YOU ARE VERY ACTIVE

Very active means that you train for around five hours over the course of a week. That can be five one-hour sessions or six fifty-minute sessions, but remember: the harder you train, the more strength training you do, the more protein you will need to include in your meals.

That's why the numbers below aren't hard and fast. If you are doing serious training, talk with your coach and your GP about what your body needs.

YOUR WEIGHT	YOUR PROTEIN REQUIREMENT
7 stone	55–77 g
8 stone	63–89 g
9 stone	71–100 g
10 stone	79–112 g
11½ stone	87–124 g
12½ stone	95–135 g

If you are lightly active and, say, 9 stone, the chart on page 59 indicates that you want to eat about 47 grams of protein each day.

Here's what to do with that number: divide it into three. Protein should be spread across the day, so that's about fifteen grams of protein per meal. Here's an example of how you might satisfy your protein needs over the course of a day.

BREAKFAST: If you have two egg whites and one yolk for breakfast, with fifteen grams of grated cheese, that's about fifteen grams of protein.

LUNCH: If lunch is a salad that includes 55 grams of chickpeas, 95 grams of brown rice, kale, tomatoes, cucumbers, parsley, and lemon juice, you'll get about eleven grams of protein.

SNACK: A handful of almonds a few hours later as a snack adds seven more grams, bringing you up to thirty-three grams by midafternoon.

DINNER: Fill your plate with 55 grams of salmon and a spinach and lentil salad, and you're at about forty-seven grams.

When I want to know exactly how much protein I'm getting, I read labels or look online at the protein values of the foods I'm eating. But honestly, after a while of eating protein regularly and consciously, including the right quantity of it in your meals will be something you just do without really having to think about it.

GET YOUR PROTEIN FIX HERE!

Luckily for you, there are so many foods that contain protein that it's not that hard to incorporate it into your meals and snacks. The chart on the opposite page lists the grams of protein found in common protein sources. You'll quickly see how easy it is to incorporate healthy sources of protein into your day!

FOOD	SERVING SIZE	PROTEIN
turkey breast	85 grams	26 g
salmon	85 grams	22 g
chicken, without the skin	85 grams	21 g
ground beef	85 grams	21 g
tuna	85 grams	20 g
cottage cheese, low fat	125 grams	13 g
edamame, frozen	155 grams	12 g
Greek yoghurt	160 grams	11 g
meatless burger, vegetable- or soy-based	1 patty (70 grams)	11 g
tofu	95 grams	10 g
lentils, cooked	100 grams	9 g
raw almonds	40 grams	7 g
low-fat (1%) milk	250ml	8 g
peanut butter	2 tablespoons	8 g
cheese (Cheddar, Colby, Brie, blue, Monterey Jack, and Swiss)	30 grams	7 g
pasta, cooked	175 grams	7 g
black beans, cooked	85 grams	7 g
egg	1 large	6 g
frankfurter, beef	1	6 g
quinoa, cooked	100 grams	4 g
broccoli, cooked	80 grams	3 g
rice, white or brown, cooked	95 grams	3 g
couscous, cooked	95 grams	3 g
whole-grain bread	1 slice (30 grams)	3 g
porridge, cooked	60 grams	3 g

COMPLETE YOUR PROTEINS

When we talk about COMPLETE proteins, we're talking about having a full stock of amino acids ready so your body can use them in whatever combination it needs. You can get your protein from animal sources or from plant sources. All of the protein you eat, no matter what the source, will be broken down into amino acids. Your body can then use those amino acids to build muscles, create enzymes or form hormones. But each of these processes requires different amino acids, so if any of them are missing, the whole operation is thrown off. When you eat complete proteins, you are giving your body *all* of the amino acids it needs.

Animal proteins are great because they are a complete-protein source. Examples include lean red meat, poultry, fish, eggs, milk and cheese. Plants are also a good source of protein because they contain amino acids, but different plants have different amino acid clusters. When you pair the right plants together, they combine to form a complete protein. For example, pairing beans or other pulses with whole grains like brown rice or quinoa offers you

AN INTERNATIONAL AMINO BUFFET

Cultures around the globe have been pairing up delicious combinations of plant-based proteins and grains for centuries. Call it nutritional intuition . . . but these simple combinations are a great way to maximize the benefits of plant-based proteins.

- **Tacos:** Mexican food combines corn tortillas with beans
- **Succotash:** Native Americans paired corn with beans
- **Sushi:** The Japanese pair rice with soy
- **Peanut stew:** West African cuisine incorporates rice and peanuts
- **Cajun red beans and rice:** The official Monday dish in New Orleans restaurants
- **Dal:** Indian lentils served over rice
- **Chana masala:** Indian chickpea dish eaten with rice
- **Mujadara:** Syrian lentil and rice stew
- **Gallo pinto:** rice and beans breakfast dish eaten in Costa Rica

Personally, eggs are one of my favourite sources of protein.

A lot of people are nervous about eating too many eggs because they've been told that eggs are high in cholesterol. Well, here's a fact: The egg **white** is pure protein. The egg yolk contains fat (and some cholesterol) as well as all of the nutrition. I mean, think about it: if that yolk became a baby chick, nutrition is what helps it grow from a cell into a chicken. As long as your doctor hasn't advised you to cut back on eggs because you have high cholesterol, don't fear this inexpensive and easy-to-prepare protein! If you're already an egg-eater, just be sure to mix in some egg whites so you're not eating too many yolks.

When I'm craving an omelette, for every three eggs, I use one yolk. That way, I get the flavour and nutrition from the yolk and an extra dose of protein. Add a little kale and Parmesan, with a side of quinoa . . . yum. The perfect protein-packed lunch or dinner!

a complete-protein source. (Check out the box on the opposite page for more great sources of complete plant-based proteins!)

THE PROTEIN POWER PACK

I appreciate protein-packed foods because they are savoury and delicious and because protein helps my body do so many things. Every time I lift a weight, do a crunch or hold a plank. Every time I pick up my niece. Every time I can haul my suitcase up the stairs by myself or help someone else with heavy bags. The protein that I eat throughout the day gives my body the resources it needs to take care of my bones and muscles, supporting me so that I can support myself and support the people around me.

I want to be as strong as possible, as healthy as possible, as continually capable as possible. Being capable is my number-one top priority, so I make sure to eat my protein from a variety of different sources, from egg whites with breakfast, to lentils with brown rice at lunch, to my favourite, whole-grain quinoa, along with chicken or fish for dinner. Protein contains the ingredients that allow my body to build and repair, and nourishes my physical self.

Protein is the source of my strength.

CHAPTER 9

FAT IS ESSENTIAL

AHHH, FATS! I LOVE them as much as the next person, maybe even more. Yummy, satisfying, nurturing, filling fats. There is so much joy in fats and in the creaminess and richness they add to a meal, so it's a good thing that there are many fats that are GOOD for us! So good for us, in fact, that fats should account for 20 to 35 per cent of our total energy intake.

I'm talking essential fatty acids, baby. Those are the fats that keep our hair and skin glowing and support the function of our organs (especially the brain and liver). Fats also provide a gentlemanly escort to the vitamins and minerals we ingest from plants, some of which aren't of any use to our cells unless they're partnered with fats. Essential fatty acids are just that, *essential* to our health.

Of course, there are also some fats that are not so good for us, including fats derived from animal protein sources and animal products (like dairy), as well as scary fats like trans fats, which are chemically engineered fats that the food industry created to extend the shelf life of snack foods (more on that later). But for now, let's focus on the fats we love to love. Remember, the right amount of fat is a necessary component of your diet! It gives you energy. It gives you essential nutrients, like vitamins A, D and E.

Yay fat!!

A SPOONFUL OF FAT HELPS THE VITAMINS GO DOWN

When we eat vitamins from plants and animals, our bodies have different ways of absorbing and using them. Some vitamins are **water soluble**, which means that they dissolve in water, and the body doesn't hold on to them for very long. B-complex vitamins and C vitamins are examples of water-soluble vitamins.

Other vitamins are **fat soluble**, which means that they can be dissolved and absorbed by the body only in the presence of fat. Think olive oil on rocket or fresh mozzarella with tomatoes. Fat-soluble vitamins can be stored for weeks or months. Vitamins A, D, E and K are fat soluble.

RE-FRIENDING FAT

Since the 1980s, fat has earned a bad rap. If you're in your twenties now, that means that during your formative years, while you were taking in information about your health and nutrition and figuring out how you'd be eating, the world around you was outfitting fat with a pitchfork and a tail, making it out to be a demon hell-bent on your destruction (or at least the destruction of your ability to fit into skinny jeans). This demonization of fat was not only misleading; it also had the unintended result of making really rich, decadent forms of fat (like cheesecake, ice cream sundaes or anything smothered in melted cheese) extra alluring – that thing that you *really really* want but you absolutely mustn't, shouldn't have.

So let's clarify what fat really is and put this issue to rest. First, I offer you five fast facts on fats and why they are good for you.

- Fat enhances the flavour and texture of food and helps all those herbs and spices and flavours really WOW your palate.
- Fat keeps your skin from being rough and scaly.
- Fat helps your body absorb vitamins.
- Fat provides you with fuel throughout the day.
- Fat boosts brainpower.

The two key things to remember about fat are that you must select the right KINDS of fat, and that you must be aware of HOW MUCH of it you're eating. When it comes to the amount of fat you eat, moderation is key. When you eat more fat than is recommended, you can increase your risk of developing heart disease and obesity. Because fats are more energy dense than proteins and carbs, a little goes a long way. Just like anything else in life, too much of a good thing can still be, well, too much.

THE FATS WE LOVE

When it comes to choosing your fats, you want to aim for the *unsaturated* variety. Unsaturated fats are liquid at room temperature. As with many other nutrients, your body needs these, but it can't make them, so it's essential to get them from the food you eat.

There are two kinds of unsaturated fats: polyunsaturated and monounsaturated. Polyunsaturated fats are primarily found in vegetable oils (such as safflower, sesame, soya, corn and sunflower) as well as nuts and seeds. These fats have a variety of benefits, from helping to protect your muscles to helping your blood to clot.

Monounsaturated fats can be found in foods like olive oil, rapeseed oil, peanut oil, avocados, and nuts. These fats are beneficial for your blood cholesterol levels and insulin and blood-sugar regulation.

You've probably heard about omega-3 fatty acids, which are among the essential fatty acids we need. Omega-3s are unsaturated fats, and they're typically found in fatty fish, like salmon, tuna and mackerel, as well as a few

OMEGA-3s FOR VEGETARIANS

Omega-3s comprise 3 kinds of fatty acids: ALA (alphalinoleic acid), EPA (eicosapentaenoic acid), and DHA (docosahexaenoic acid).

- Fatty fish and fish-oil products provide EPA and DHA, but these sources may not be appropriate for a vegan or an ovo-lacto vegetarian.
- Ground linseed, walnuts, soya beans, soya oil, rapeseed oil and marine algae are all examples of plant-based sources of omega-3s.

As with essential amino acids, our bodies cannot produce these essential fatty acids and we cannot live without them, so we must get them from our food sources.

plant-based sources (check out the box on the opposite page). You can believe the hype about these much-buzzed-about fats: they are superstars! From protecting you against heart disease and Alzheimer's to boosting your brainpower, omega-3s are some of the best fats you can choose.

THE FATS WE LIMIT

When we're choosing our fats, we want to limit the amount of *saturated* fats, *trans* fats and cholesterol that we select. These fats are the ones that may bear a threat to the health of our arteries and are the ones usually found in fast food and processed foods, as well as in dairy products, such as butter, cheese and milk, and meat products, as well as in coconut oil and palm kernel oil (found in many processed sweets). There are about twenty-four different saturated fats, and not all of them are bad for you. For example, coconut oil has saturated fat, but it also helps boost your 'good' cholesterol level and your thyroid function.

When I indulge in saturated fats, it's usually in the form of a nice piece of Cuban pork or a really delicious burger. But I don't eat that way every day. And things like fast-food burgers? Well, every once in a while I'll get an urge and I'll go ahead and give in. I get the smallest burger and fries available and enjoy the taste for that moment. But I always know that within the next thirty minutes, I'll be nursing my poor belly, which is the one that takes the hard fall in order for my mouth to enjoy that taste of nostalgia. But that's not something I do very often. If I'm going to eat a burger, I prefer to either make it myself or go to a restaurant or burger joint that makes their burgers from fresh ingredients. High-quality fresh meats and cheeses still offer your body some nutrition, but fast food is just empty calories.

Speaking of natural versus artificial, there is NOTHING natural about trans fats. They are literally man-made fats, and they are no good. All natural foods spoil at some point, so food manufacturers figured out a way to create a fat that could keep their products 'fresh' for months on end. Basically, they add hydrogen molecules to vegetable oil in order to create this extremely shelf-stable, solid fat. Margarine and vegetable shortening contain trans fats, and both of those ingredients are used in a ton of fast foods and processed foods, especially the kinds of foods you find at convenience stores and petrol stations (crisps, cakes, biscuits, crackers, etc.). These fats have no saving graces. There is NO acceptable amount of trans fat.

COOKING WITH FAT

At home, I cook with olive oil when I'm making roasted vegetables and chicken, and I drizzle olive oil over avocado and Parmesan. When it comes to frying, I love grapeseed oil. Which oils should you use? That's a matter of taste – and of smoke point.

Oils can have low, medium or high smoke points. Smoke point is the temperature at which an oil starts to smoke. The smoke point tells us how an oil can handle heat. Oils with *low* smoke points can't handle the heat, so they're best reserved for dressings and dips. Oils with *medium* smoke points are ideal for everyday cooking on the hob and baking. Oils with *high* smoke points can withstand very high temperatures, so you can really turn up the fire and flash-fry or sear a piece of fish or meat.

LOW SMOKE POINT OILS

Best used in: Salad dressings, marinades and dips

- **Walnut oil:** heart-healthy, full-flavoured and delicious drizzled over vegetables or used to dress your salads
- **Flax Seed oil:** like walnut oil, ideal to use in salad dressings or mixed into a smoothie; a great source of omega-3s
- **Extra-virgin olive oil:** processed rapidly after the olives are picked, very flavourful, and great as a finishing oil or dip

MEDIUM SMOKE POINT OILS

Best used in: Sautéing, sauce making, stir-frying and oven baking

- **Olive oil:** my general, all around go-to choice – perfect for cooking proteins or veggies
- **Rapeseed oil:** versatile and a great source of monounsaturated fats;

has a light flavour that works well for baking and can also be used to coat your barbecue grill

- **Coconut oil:** gives foods a delightful, light coconut flavour; works well for curries and sautéing tofu
- **Grapeseed oil:** extracted from grape seeds during wine making; earthy flavour, a great choice for sautéing
- **Sesame oil:** a delicious nutty flavour that adds a nice depth to Asian-style dishes

HIGH SMOKE POINT OILS

Best used in: Searing, browning and pan-frying

- **Safflower oil:** a good source of vitamin E, with a mild flavour; can be used in everything from curries to baking
- **Sunflower oil:** full of vitamins A, D, E; a good choice for frying
- **Groundnut oil:** monounsaturated, contains essential fatty acids, really lends a peanut flavour to food; another top choice for frying

CHOOSE WISELY!

Think of fats as being like a group of people at a party. Some people are wonderful and well worth your time and some are jerks, and it's your job to judge each of them on merit. While some fats – the saturated and trans fats – are like the alluring but ultimately destructive guy you really shouldn't date, others are like the nice guy next door that your best friends keep wishing (rightly) that you would fall for.

So olive oil, yes! I'd love to spend more time with you. Margarine, no, I'm afraid I'm busy washing my hair on Friday night. With so many heart-healthy fats out there to try, you'll always have something delicious and full of vitamins and minerals to pour, whisk and drizzle.

CHAPTER 10

EATING THE STARS

————

W HEN I COOK MY protein-rich eggs in an iron frying pan coated with olive oil and pair them with the wonderfully healthy fats of a sliced avocado, some complex carbs in the form of sprouted rye bread and all of the nutrients of a bright, juicy orange, I'm getting a delicious breakfast.

I'm also getting a full course of vitamins and minerals, including some of the iron from the pan, which gets absorbed into my eggs in trace amounts. In addition, my eggs contain their own iron as well as magnesium and calcium. My avocado contains potassium, phosphorus and zinc, plus a bunch more vitamins.

When you eat plants and animals that eat plants, not only are you eating the sun, you are also eating the stars. The calcium, magnesium and iron that are soaked up from the soil by plants that are then eaten by animals and people are the same minerals you'll find in stars in the sky. They're also the same minerals you'll find in us humans here on the ground (our bodies are 4 per cent minerals!).

Most food is or recently was alive, so most nutrients, like vitamins, are organic. Minerals are not alive. They are called *inorganic* substances to distinguish them from the more complex, carbon-based *organic* materials that make up living things. And all of these vitamins and minerals, as you'll see, are absolutely necessary at the right levels to keep us healthy. Vitamins and minerals are so good at their jobs that there are a host of diseases you've probably never even heard of that you're being protected against daily when you eat apples and bananas and spinach.

WHAT HAPPENS WHEN YOU DON'T EAT THE STARS

Here's a brief history lesson: one hundred years ago, the Southern United States was struck by a mysterious plague. They called the disease pellagra. Sufferers' skin changed texture and became thick and strange. They lost their minds. And many people died.

By 1914, one hundred thousand people were sick. But nobody knew why! The assumption was that it was a contagious disease, but doctors were unable to offer conclusive proof – until a man named Joseph Goldberger got involved. He went down South and started asking pellagra victims questions and paying attention to the answers. Finally, Goldberger thought he had it. Through his conversations with the sick, he knew that most pellagra victims were poor and getting by on a restricted diet that consisted mostly of cornbread, molasses and some pork fat. They ate no, or very few, fresh fruits and vegetables.

The cause of the disease, he deduced, was poor nutrition.

Do you love having soft, smooth skin? Do you love being able to think clearly? To see clearly? Then eat your fruit and vegetables!!

In order to get people to believe that pellagra was the result of a low-quality diet and not a contagion, Goldberger conducted an experiment. He found a farm-based prison where the inmates were healthy and ate plenty of produce. He changed their diets to mimic what the pellagra sufferers had been eating. Within months, all of the inmates had come down with pellagra. When they were given fresh fruit and vegetables again, their health improved!

In the end, Goldberger's findings were confirmed. Pellagra wasn't contagious; it was the result of a nutritional deficiency. He died before the actual deficient nutrient – niacin, vitamin B_3 – was discovered, but he was right about the power of fruits and vegetables. The absence of niacin affected people's skin. It affected their brains. And it ended their lives.

Even though niacin hadn't yet been identified as a vitamin, it was still clear that fruit and vegetables possessed the ability to heal. Which is probably why my parents always told me to eat my fruit and vegetables. . . .

FRUIT + VEG = MICRONUTRIENTS

Ever since I was a kid, people have been urging me towards the salad bar. I bet it's the same for you. Perhaps you listened and devoured those platefuls of brussels sprouts. Or maybe your dog had a steady diet of green side dishes, served under the table while your parents couldn't see. Either way, now that you're an adult, eating your fruits and vegetables should be a part of your regular routine.

Do you love having soft, smooth skin? Do you love being able to think clearly? To see clearly? Then eat your fruit and vegetables!!

The more I learn about vitamins and minerals, the more I realize how lucky I am to live in a place where fruit and vegetables are available to me year-round. Simply eating fresh fruit and vegetables gives me heaps of calcium for my bones, iron for my blood and vitamin C for my immune system. And whether I'm on the road, doing my training, or memorizing lines, all those spinach salads, rocket salads, bowls of cherries, and plates full of broccoli, corn, aubergine and tomatoes make me feel sharper, stronger and more capable.

WHAT YOU HAVE IN COMMON WITH PLANTS

Just as our bodies store sugar as carbs and fats, plants store extra glucose as starches and fats. Think of sweet vegetables, like carrots and beetroot – that sweetness comes from the stored sugar. On the other side of the spectrum, there are creamy avocados and coconuts, examples of plants storing extra energy as fats.

THE BONE BUILDERS

calcium, vitamin D, phosphorus, magnesium

In my lifetime, with all the physical activity that I do, I'm lucky to have broken only two bones. Well, not including my nose, which I've broken four times. But I only had to break a bone once to know how important it is to make sure that I get plenty of bone builders, especially as I get older. As you will learn in great detail in Chapter 18, our bones are in a constant makeover montage, with the old cells continuously being lost and new cells being created. So supplying the nutrients our bodies need to build healthy bones – calcium, vitamin D, phosphorous and magnesium – is integral if you want to maintain your bones as an adult. If you eat whole foods, you probably get enough phosphorus and magnesium, but many women do not get adequate calcium from food and may want to consider a calcium supplement.

SURPRISING SOURCES OF CALCIUM

When people think of getting enough calcium, they usually turn to dairy foods. But dark green leafy vegetables (broccoli, kale, pak choi, turnip and spring greens) and other nondairy foods also offer a nice dose of calcium as well. So instead of a glass of milk, how about . . .

- soya milk, 250ml: 300 mg (as much calcium as cow's milk!)
- soya beans, 180 grams, cooked: 261 mg
- broccoli, 90 grams, cooked: 180 mg
- white beans, 130 grams, cooked: 100 mg
- kale, 70 grams, raw: 90 mg
- almonds, 30 grams: 80 mg

THE BONE BUILDERS

NUTRIENT	WHAT IT DOES...	THE MAGIC NUMBER [a,b,c,d]	WHAT DEFICIENCY LOOKS LIKE	WHERE YOU GET IT
CALCIUM	Bone structure Muscle contraction Blood clotting Nerve impulse transmission Secretion of hormones	700 mg/day but not more than 1500 mg/day	Stunted growth (childhood); reduced bone mass (adults); osteoporosis (in advanced age)	Milk and milk products, green vegetables, pulses, tofu, fish (with bones)
PHOSPHORUS	Structure of bone and teeth pH balance in the body Energy reactions Helps form cell membranes and genetic material	250 mg/day but not more than 550 mg/day	Muscle weakness, bone pain; rarely occurs from poor diet; can occur with alcohol abuse and medications that bind phosphorus	Meat, fish, poultry, eggs, milk and milk products
MAGNESIUM	Structure of bone and teeth Helps the body make proteins Muscle contraction Blood clotting	270 mg/day but not more than 350 mg/day[e]	Muscle weakness, confusion, stunted growth in children	Green leafy vegetables, whole grains, nuts, seeds, seafood, beans, chocolate, cocoa
VITAMIN D	Healthy bones Regulates blood calcium levels Serotonin production	15 mcg/day but not more than 25 mcg/day	Rickets in children (weak bones, bowed legs), osteomalacia in adults (soft, brittle bones)	Fortified milk, eggs, butter, fatty fish (salmon, sardines, herring), synthesized in the body after exposure to sunlight

[a] Dietary Reference Intakes for women 19–30 years.

[b] NHS RDA = Recommended Daily Allowances (the average daily intake level sufficient to meet requirements in the body).

[c] AI = Adequate Intake (provided instead of an RDA when scientific evidence is not available).

[d] UL = Tolerable Upper Intake Level (the maximum daily intake level unlikely to cause adverse health effects).

[e] The UL for magnesium applies only to intake from dietary supplements or pharmacological supplements and does not include intake from food and water.

VITAMIN HAPPY

Moods are sometimes related to micronutrients. For instance, vitamin D plays a role in the production of serotonin, a hormone that promotes positive feelings in our brains. A deficiency of D is associated with lousy moods and lack of energy, which means that not getting enough vitamin D means not getting enough happy.

Lucky humans, we can eat our vitamin D – or get it from the sunshine. While we can't get our energy directly from the sun's glittering rays, we can get our daily dose of D, because our bodies can make vitamin D from sunlight. A quick jog outside in the morning, and I've not only recharged and had a chance to sweat, I've received some vitamin D, too!

The same way that plants use light to create their fuel via photosynthesis, your skin uses the light of the sun to photosynthesize vitamin D. At midday in the summertime, it takes only twenty minutes of sun exposure for the body to make 20,000 IU (international units) of vitamin D . . . and when you consider that the recommended daily allowance for people under the age of fifty is 200 IU, it's clear that the sun is an effective supplier (as long as it's not raining).

You can eat your vitamin D too, of course, just in case you're trapped in a monsoon. Foods like fortified milk, eggs, butter and fatty fish are all great sources of vitamin D. If you suspect you're not getting enough D, your doctor can give you a simple test. And if you have trouble fitting in enough time in the sun or D-centric foods, you can also take a supplement.

But remember, once you've soaked up your dose of D, you've got to protect your skin. Be sure to apply sunscreen on all exposed parts of your body if you plan to be in the sun longer than twenty minutes.

THE BLOOD FORMERS

iron, copper, folate, B$_{12}$

Think of the blood that courses through your veins as a busy river – it's used to transport a lot of different things to various far-off places. Your blood delivers oxygen throughout your body, brings carbon dioxide back to the lungs, delivers nutrients to your cells and transports waste products to be eliminated from the body. Your body needs iron, copper, folate and B$_{12}$ so that it can make healthy red blood cells to do all of these jobs and more.

The most common nutrient deficiency in the world is iron deficiency, which causes anaemia – and women are more likely than men to become anaemic. The most common causes of an iron deficiency are blood loss, not getting enough iron in your diet or having an intestinal disorder, like coeliac disease, that prevents you properly absorbing iron. Pregnant women, vegetarians and women with very heavy periods are at a higher risk for developing anaemia, whose symptoms can include tiredness, dizziness, shortness of breath, cold hands and feet and rapid heartbeat. If you feel that you might be

VEGETARIANS: PUMPING IRON

If you are a vegetarian, you should be consuming almost double the recommended daily intake of iron to prevent a deficiency. The best plant sources of iron include pulses; soya beans; tofu; nuts; dried fruit, like apricots; dark green leafy vegetables, like spinach, spring greens, kale and turnip greens; and fortified cereals. But veggie lovers have to take extra precautions. The body isn't that great at absorbing iron from plant sources, but vitamin C helps with that process. So squeeze some lemon juice on your steamed kale, make a beautiful salad with spinach and strawberries or top your tofu with stewed tomatoes (see page 79 for a list of foods that contain vitamin C). Always try to pair your iron with a vitamin C–rich food for maximum nutritional benefit.

Getting enough B$_{12}$ can also be a struggle for vegetarians, because it is found only in animal products. Vitamin B$_{12}$ is responsible for some very important functions, like maintaining your nerve tissue; if you go without it for too long, you can ultimately damage your nervous system. Vegetarians should look for B$_{12}$-fortified products, like cereals and nutritional yeast, and may want to consider supplementation as well. Your GP can also give you a B$_{12}$ shot.

THE BLOOD FORMERS

NUTRIENT	WHAT IT DOES...	THE MAGIC NUMBER [a,b,c,d]	WHERE YOU GET IT	WHAT DEFICIENCY LOOKS LIKE
FOLATE	DNA and red blood cells Amino acid metabolism	0.2 mg/day but not more than 1mg/day[e]	Leafy green vegetables, citrus fruits/juices, organ meats, pulses, seeds, fortified cereals and grains	Anaemia, weakness, fatigue, headache, difficulty concentrating, sores in mouth and on the tongue, increased risk of giving birth to an infant with a neural tube defect
VITAMIN B$_{12}$	Helps you use folate properly Fat and amino acid breakdown Helps maintain nerve tissue	0.0015 mg/day but not more than 2mg	Meat, fish, poultry, milk, fortified cereals and grains; found only in animal products	Anaemia, fatigue, impaired short-term memory, nerve damage leading to paralysis
IRON	Forms haemoglobin, the oxygen-carrying protein of the red blood cells Forms myoglobin, the oxygen-carrying protein of muscle	14.8 mg/day but not more than 20 mg/day	Meat, fish, poultry, whole grains, eggs, pulses, dried fruits, fortified cereals and grains	Anaemia, weakness, fatigue, headache, pale skin, poor resistance to cold temperatures, decreased ability to exercise, poor cognitive function
COPPER	Helps your body use iron Defends the body from unstable molecules	Not more than 1 mg/day	Offal, seafood, nuts, seeds, whole grains	Anaemia, bone abnormalities

[a] Dietary Reference Intakes for women 19–50 years.

[b] NHS RDA = Recommended Daily Allowances (the average daily intake level sufficient to meet requirements in the body).

[c] AI = Adequate Intake (provided instead of an RDA when scientific evidence is not available).

[d] UL = Tolerable Upper Intake Level (the maximum daily intake level unlikely to cause adverse health effects).

[e] The UL for folate applies only to synthetic forms obtained from vitamin supplements and fortified foods.

anaemic, see your GP. She can put you on an iron supplement if necessary, but don't try supplementing on your own without a diagnosis – too much iron in your body puts a strain on your liver.

Women who are thinking of having children will want to consider how much folate they are getting in their diets. Many women supplement with folic acid before and during pregnancy to prevent birth defects.

THE ANTIOXIDANTS

vitamin C, vitamin A, selenium, beta-carotene, vitamin E

Sometimes I like to have a snack of apple slices with almond butter. If I slice the apple and then go off to answer the phone, when I come back, inevitably, that crisp white Pink Lady apple has started to look a little bit . . . brown.

That browning process is the result of something called oxidation. What does the flesh of an apple have to do with the flesh of your body? Well, when you are exposed to secondhand smoke or polluted air, a similar process happens inside of you. In the case of the apple's cells, no big deal. But when it's *your* cells, you want to pay a little bit more attention. Sure, your body is constantly regenerating cells, so while the apple gets brown and rotten, you get new cells to replace the old ones. But sometimes in the damaged cells, molecules get 'dinged', losing an electron or two and becoming what are called free radicals. Missing something of themselves, so to speak, they go out to reclaim what is lost, sometimes damaging other cells and initiating a chain of cellular events that can lead to disease . . . like heart disease, cancer, arthritis, cataracts and diabetes. Free radicals continue the oxidation at a cellular level, hurting others as they have been hurt, and in essence ageing the body (including your skin).

So what can you do about it? You can help your body by supplying it with liberal amounts of *ANTI*oxidants in the form of oranges, carrots, green leafy vegetables, whole grains, strawberries, nuts . . . the very same foods we've been talking about all along.

Because vitamin C, vitamin A, selenium, beta-carotene and vitamin E are part of the antioxidant army that protects your body from damage. Antioxidants, as the name implies, protect against oxidation.

THE ANTIOXIDANT ARMY

NUTRIENT	WHAT IT DOES ...	THE MAGIC NUMBER [a,b,c,d]	WHERE YOU GET IT	WHAT DEFICIENCY LOOKS LIKE
VITAMIN C	Helps make collagen Immune system support Helps you absorb iron	40 mg/day but not more than 1000 mg/day	Citrus fruits, dark green vegetables, potatoes, muskmelon, strawberries, tomatoes	Scurvy (bleeding gums, pinpoint bleeding, abnormal bone growth, bone pain), poor wound healing, anaemia, depression
VITAMIN E	Immune system booster Protects vitamin A and polyunsaturated fatty acids from oxidation	3 mg/day but not more than 540 mg/day[e]	Vegetable oils, salad dressing, margarine, nuts, seeds, green leafy vegetables	Red blood cell breakage, anaemia, nerve damage, muscle weakness, muscle degeneration, fibrocystic breast disease
SELENIUM	Protects cell membranes Immune system support Regulates thyroid function	0.35 mg/day	Offal, seafood, whole grains, meat, vegetables	Keshan disease (a form of heart disease), Kashin-Beck disease (a form of arthritis), impaired immunity
VITAMIN A	Helps your eyes adjust to changes in light Reproduction Bone growth	0.6mg/day[f] but not more than 1.5mg/day	Fortified milk, cheese, cream, butter, eggs, liver	Night blindness, impaired immune function, degeneration of cornea leading to blindness, hair loss
BETA-CAROTENE	Protects cell membranes Protects eyes Immune system support	No more than 7mg/day	Spinach and other leafy greens, broccoli, carrots, apricots, muskmelon, sweet potatoes, pumpkin	Not known

[a] Dietary Reference Intakes for women 19–50 years.

[b] NHS RDA = Recommended Daily Allowances (the average daily intake level sufficient to meet requirements in the body).

[c] AI = Adequate Intake (provided instead of an RDA when scientific evidence is not available).

[d] UL = Tolerable Upper Intake Level (the maximum daily intake level unlikely to cause adverse health effects).

[e] The UL for vitamin E (as α-tocopherol) applies only to synthetic forms obtained from dietary supplements, fortified foods, or a combination of both.

[f] As retinol activity equivalents (RAEs). 1 RAE = 1 μg retinol, 12 μg ß-carotene, 24 μg α-carotene, or 24 μg ß-cryptoxanthin.

For instance, vitamin E hangs out in your cell membranes. When a free radical tries to damage that membrane, vitamin E takes the blow instead. Vitamin C helps your body absorb iron, boosts your immune system and plays an important role in maintaining healthy skin. Selenium is also great for your immune system, and so is beta-carotene. An antioxidant, beta-carotene protects your eyes and cell membranes, and the body converts it to vitamin A, which helps your eyes adjust to changes in light and promotes good vision.

Eating an antioxidant-rich diet is one of the most important things we can do to protect ourselves from the onset of disease as well as to look and feel healthier longer.

THE ENERGY VITAMINS

thiamin, riboflavin, niacin, and vitamin B_6

I am a busy lady, and I need a lot of energy to keep up with everything I have booked on my calendar. I'm always on another plane, or at another meeting, filming, training, running errands, grocery shopping or cooking. There's so much to see and do and enjoy and experience, and I need maximum energy if I'm going to accomplish the maximum.

We've already talked about how energy comes from the macronutrients – carbs, protein and fat. But they don't do it alone: micronutrients play a role in that process too, especially the B vitamins. Thiamin, riboflavin, niacin and B_6 are four of the eight B vitamins, used by the body to help convert carbs into glucose, which is used to produce energy. B vitamins are the Energizer Bunnies of the micronutrients, helping you to get physical and perform optimally.

Thiamine, or vitamin B_1, helps your body break down sugars and is important for a healthy nervous system. Riboflavin, or B_2, must be consumed every day. It helps your body make red blood cells and it is important for your skin, nails, and hair. Niacin helps to improve circulation and aids the body in producing stress hormones and sex hormones. And B_6 works to fight infections, keep blood sugar levels regular and build haemoglobin for red blood cells.

THE ENERGY VITAMINS

NUTRIENT	WHAT IT DOES...	THE MAGIC NUMBER [a,b,c,d]	WHERE YOU GET IT	WHAT DEFICIENCY LOOKS LIKE
THIAMIN	Helps you get energy from carbs and proteins Helps your nerves send messages	0.8 mg/day	Whole-grain products, pork, ham, liver, dark green vegetables, nuts	Apathy, less short-term memory, confusion, irritability, muscle weakness, damage to heart tissue
RIBOFLAVIN	Helps you metabolize energy from food Helps you metabolize other vitamins (folate, vitamin B_6, niacin)	1.1 mg/day	Milk and dairy products, whole-grain products, offal	Inflammation of the mouth, skin, and eyes, sore throat, magenta tongue, cracks at the corner of the mouth
NIACIN	Helps turn carbs, fat, and alcohol into energy	13 mg/day but not more than 17 mg[e]	Brewer's yeast, meat, fish, poultry, mushrooms, nuts, pulses, whole grains	Pellagra: diarrhoea, vomiting, depression, fatigue, rash on areas exposed to sunlight
VITAMIN B_6	Helps metabolize amino acids Helps your body break down glycogen	1.2 mg/day but not more than 10 mg	Animal foods, such as meats, fish and poultry, fortified cereals and grains, noncitrus fruits, vegetables	Dermatitis, small-cell anaemia, depression, confusion, convulsions

[a] Dietary Reference Intakes for women 19–50 years.

[b] NHS RDA = Recommended Daily Allowances (the average daily intake level sufficient to meet requirements in the body).

[c] AI = Adequate Intake (provided instead of an RDA when scientific evidence is not available).

[d] UL = Tolerable Upper Intake Level (the maximum daily intake level unlikely to cause adverse health effects).

[e] The UL for niacin applies only to synthetic forms obtained from vitamin supplements and fortified foods.

This is one case where the B team is of A-team importance. These guys are your digestion champions, your energy crew. Every time I go dancing, go swimming, or get my sweat on it's the B team supporting me. Want to have more energy, a good memory and high spirits? Then you need plenty of leafy green vegetables, whole grains, fish, mushrooms and high-quality dairy products.

THE HYDRATERS

sodium, potassium, and chloride

I'm a salt lover and a water lover, and lucky for me, those things go together beautifully. Sodium, which is half of what salt is, helps maintain fluid balance in the body. Too much salt makes you feel bloated because an excess of sodium causes your body to retain water. But in the right amount, sodium is integral to our health, because it is one of the electrolytes.

You've probably seen various waters and sports drinks with labels boasting of added electrolytes. So what are electrolytes? Are they something your body needs or something marketers need you to believe in so you'll part with a few more quid? Believe it or not, they're real and important. Electrolytes are micronutrients like sodium, potassium and chloride, compounds that assist with fluid balance in the body, helping to ensure that the fluids you take in (like water) are distributed properly inside and outside of your cells. Electrolytes also help transmit nerve impulses in the body, sending signals so that your muscles can contract and you can do things like go for a walk or open a jar.

Keep in mind that sodium is important for your health, but too much is associated with hypertension, an illness that elevates blood pressure in the arteries, making the heart work much harder than it should. That's why eating too much salt is bad for you! On the flip side, eating plenty of potassium-rich foods, like bananas, may offer protection against hypertension. As with other areas of your nutrition, too little is not enough, and too much is way more than your body can handle. So, balance, balance, balance.

HOLD THE SODIUM

Seventy-seven per cent of the average person's sodium intake comes from processed or restaurant foods.

THE HYDRATING ELECTROLYTES

NUTRIENT	WHAT IT DOES ...	THE MAGIC NUMBER [a,b,c,d]	WHERE YOU GET IT	WHAT DEFICIENCY LOOKS LIKE
SODIUM	Keeps fluids in balance outside cells Helps nerves transmit information Muscle contraction	6g/day	Table salt, soya sauce	Muscle cramps, headache, dizziness, fatigue, loss of appetite, mental apathy
POTASSIUM	Keeps fluids in balance inside cells Helps nerves transmit information Muscle contraction Good for blood pressure	3500 mg/day but not more than 3700 mg/day	Fruits, vegetables, meats, grains, pulses	Muscle weakness, confusion, loss of appetite
CHLORIDE	Plays a role in fluid balance Helps form stomach acid	2300 mg/day but not more than 3600 mg/ day	Table salt, soya sauce, meats, milks, eggs	Does not typically occur

[a] Dietary Reference Intakes for women 19–50 years.

[b] NHS RDA = Recommended Daily Allowances (the average daily intake level sufficient to meet requirements in the body).

[c] AI = Adequate Intake (provided instead of an RDA when scientific evidence is not available).

[d] UL = Tolerable Upper Intake Level (the maximum daily intake level unlikely to cause adverse health effects).

I love intensity when it comes to food. The sharp taste of garlic, the vibrant red of tomatoes, the bitter tang of rocket. And, luckily for me, the same things that add those layers of flavour and colour are also really, really good for me. Garlic's bite, tomatoes' hue, rocket's spice: all of these qualities come from *phytochemicals*, beneficial chemical compounds that occur naturally in plants. How incredible that the same things that make natural foods so gorgeous to look at and so delicious to eat are the substances that help your body fight disease! For instance, lycopene, which makes tomatoes red, promotes heart health. Allicin, found in garlic, has antimicrobial properties. And it's thought that indoles found in rocket may even help fight cancer.

So feel free to use colour as an indicator of nutrient-density for your fruits and vegetables. The deeper green a leaf of lettuce is, the more nutrients it is likely to hold. When you're pulling together the ingredients for your salad, remember that it's a great opportunity to load up on micronutrients. Try chopping up raw broccoli, spinach, green leaf lettuce, red cabbage and tomatoes into bite-sized pieces. Maybe top it up with sliced avocado, and some raw sunflower seeds or chickpeas or black beans for protein. Serve it over brown rice for some complex carbs. Squeeze some lemon over the top, add a nice drizzle of olive oil to help you absorb all of those fat-soluble vitamins and a pinch of salt and pepper, and you've got yourself an *excellent*, phytonutrient-rich meal.

Don't be afraid to mix it up! The more the merrier. Vegetables love to get together, and they have an amazing ability to pair well not only inside your body but also inside your mouth. Experiment with different combinations of vegetables and consider adding fruit. Try spinach and strawberries and almonds and shallots. Or rocket with black beans, red onion and mango. Or add dried fruit, like sultanas or currants, for a hit of sweetness.

Whatever your favourite vegetable, fruit, grain and bean combination, the best dressing is always the easiest and simplest. One freshly squeezed lemon and a nice drizzle of olive oil with a pinch of salt, and if you like pepper, add a little pepper. Toss it all together and you will be in heaven! It's so simple. If you want to get more adventurous, you can experiment with adding different herbs and spices, like finely chopped fresh garlic, fresh basil or dried oregano, or chilli flakes. The point is to just play a little and try to make something that combines lots of different phytonutrients and also tastes good to you.

When I was cooking with my nieces recently, one of them asked me, "Aunty, how do you know how to make this food taste so good?"

And I said, "Because I just tried. I thought about what I wanted to eat, I looked up a recipe on the Internet and found out how to make it, and I just kept cooking it until I finally got it to taste like what I wanted it to taste like."

She said, "Oh, that makes sense."

That's the whole secret to feeding yourself well: taking the time to create food that you love to eat while also making sure that it provides you with the vitamins, minerals, carbohydrates, essential fats and protein that you need to be healthy.

The following list is a handy chart of foods that are high in vitamins, minerals and phytochemicals. As a rule of thumb, the more colourful your plate, the wider the range of nutrients.

VEGETABLES		FRUIT	BEANS	NUTS AND SEEDS	WHOLE GRAINS
asparagus	mushrooms	apples	split peas	golden linseed	brown rice
aubergine	pak choi	blackberries	soya beans	sunflower seeds	steel-cut oats
broccoli	parsley	blueberries	black beans	walnuts	quinoa
brussels sprouts	peppers	raspberries			
cauliflower	rocket	red grapefruit			
celery	spinach	red grapes			
chilli peppers	spring greens	strawberries			
courgettes	squash	watermelon			
dandelion	tomatoes				
kale	watercress				

So fill your plate with leafy greens, chewy whole grains, crunchy nuts and seeds, yellow peppers, purple aubergine, pink watermelons and red raspberries. That rainbow on your plate means your health is golden.

And if you feel you may have a deficiency, or if you have any questions about the best way to supplement your diet, don't hesitate to talk to your GP. That's what she's there for! Curiosity might have killed the cat, but the desire to know WHAT and WHY keeps women HEALTHY.

CHAPTER 11

WATER IS LIFE

VERY NIGHT BEFORE I go to sleep, I fill up a big glass bottle with water and put it on my bathroom counter. First thing in the morning, right after I brush my teeth, I drink it because I know that during the night, while I am sleeping, all of that breathing in and out exhales a lot of moisture from my lungs. Imagine how much water you can lose during eight hours of breathing! We can't replenish it while we are sleeping, so it's important to rehydrate once we wake.

Once I drink the water in the bottle by my sink, I feel it immediately. I go from being a wilted plant to one that has just been rejuvenated by the rain. All of my cells fill up with water and I become bright and vibrant again. My eyes and nose become moist, my throat doesn't feel dry and scratchy anymore, and best of all, the water kick-starts my digestive tract. After the whole bottle is empty, I refill it and drink it again over the course of the morning. I drink it at room temperature, sometimes with a lemon in it, often plain. I refill it several times throughout the day, and I make sure that I always have it close. That steady and consistent intake of water throughout the day helps me to keep my mind clear and my body in motion, not to mention that it helps me keep a smile on my face, because when your mouth is dry, it's hard for your lips to glide open and reveal those pearly whites.

But that first bottle . . . that first bottle of water in the morning is my wake-up call. I just chug the whole thing. By leaving it on the counter, I don't even have to think about it: I can just wake up and drink, drink, drink, until my whole system is awake and I feel nourished and rehydrated, instead of as gravelly and dry as the Sahara.

Water is life. Your life. My life. All of the life on this planet. Such a simple recipe – two hydrogens bonded with one oxygen – and you get the stuff that comprises more than half your body weight. That's right! 50 to 70 per cent of you, more than half, is water! Most of that water – two-thirds of it – is found within your cells. The other third is in your blood vessels and between the cells, tissues and organs of your body. And without that water, you couldn't live for very long, making finding water sources a number-one priority for all humans throughout the ages. No matter where we are born, access to fresh, clean water is one of the most important parts of our day, every single day. It's something that you and I don't even have to think about, but sadly, a lot of other people around the world do.

There's a lot of blue on the globe, but that doesn't mean there's a surplus of drinking water. Only 1 per cent – that's one out of one hundred, people – is the fresh, drinkable kind of water. Most of the water you see on the globe will never make it into your glass.

The top three survival needs for humans are food, shelter and WATER. We need all of the nutrients to stay alive, but water may be the most essential nutrient of all. If you were stuck in a place where there was rain but no food, you could probably live for about a month, which is how long the human body can survive without eating. If you were stuck in a desert where there was no water, you'd barely last a week. *Water is more important to your survival than food.*

MONITOR YOUR HYDRATION

- **Check your thirst:** Feeling thirsty doesn't just mean that you need water. It means you've needed it for a while. Thirst is your body's way of saying that it's been toooo long.
- **Check your pee:** What colour is your pee when you wake up and throughout the day? Generally speaking, a pale yellow colour means that you are well hydrated. A dark yellow colour and infrequent urination suggest dehydration.

HOW MUCH DO YOU NEED?

Your intake of water should match the losses of water from your body every day (via sweating, peeing, etc.). When you lose too much and you haven't had enough to drink, you can become dehydrated. Elevated thirst, dry mouth and skin, a rapid heartbeat, weakness and confusion are all signs of dehydration. Remember when we talked about water as a temperature regulator? Well, when you become very dehydrated, your body temperature can rise to dangerous levels, which can hurt your kidneys and cause other organ damage.

One of the sure signs of early dehydration is a heavy, foggy headache, the kind that makes it hard to think straight. Instead of reaching for the ibuprofen, try drinking a glass or two of water first, then see if that clears your head and thoughts. If it does, it won't hurt to drink another glass for good measure, because hydration is essential to every part of your body.

Your body needs about two litres of water a day to stay hydrated. The precise amount of water you need on a daily basis will depend on your age, your health, your activity level, your environment (you probably need to drink more water if you live in Arizona than if you live in Alaska) and your diet – fruit and vegetables are typically more than 90 per cent water, so when you eat a juicy peach or a cucumber and tomato salad, you are absorbing the water that the fruit and vegetables are holding. Hydration is one more benefit of eating those delicious, juicy foods, and another great example of 'you are what you eat'.

Remember, your best source of water is water. Not energy drinks. Not iced tea or lemonade. Not coffee, not alcohol, not juice, and definitely not a fizzy drink. When you think, 'I need a drink', make that drink *water*. If you love your beverages and you want something more than just plain water, squeeze a little lemon or orange into your glass or smash up some berries and throw them into your bottle of water. Its fun to play with fruits to see what flavour blends you can come up with. Plus, you'll get some of the nutrients from the fruit.

That first drink of water I have in the morning? It invigorates my entire body. As soon as that litre of water is down, I can feel it working. I can feel all of my cells filling up like a blossoming flower.

CHAPTER 12

THE DIGESTION CHAPTER

YOU ARE A LOT like an earthworm. Don't take it personally: so am I. We may get manicures and pedicures and thrill at new shoes, but when we pour water and food into our mouths, we are starting them along a route that is basically one long tube designed to ingest food, process food and get rid of food. As your grandmother probably used to say, "In one end, out the other."

In between the ingesting end and the getting-rid-of end is that tube, which is *nine metres* long, all coiled up to fit inside your body – an incredible feat, as I'm guessing your body is not nine metres tall. Along that tube, your body has an amazingly organized team of efficient workers that convert your food into fuel, a process that takes up to seventy-two hours (depending, of course, on your body type, how much you've eaten and what you've eaten).

According to Dr Alejandro Junger, author of *Clean Gut*, your digestive system is the epicentre of your health. It is where your body extracts nutrition, and it is part of your body's immune defence system. It has the massive responsibility of processing what you eat and extracting the nutrients so that they can be distributed throughout your body. The healthier your digestive system, the better it can process food, break the nutrients down into smaller units and absorb that nutrition so it can be sent to your cells. Anything that doesn't get digested and absorbed – including toxins that can make you ill – is eventually eliminated from your body.

Digestion begins in your mouth. The very same mouth that talks a good game, is great at kissing and looks fabulous in red lipstick is also a handy tool for digestion. You've got those pretty, white teeth, which tear and cut food into manageable pieces. Chewing gets your digestion started: all that gnashing and gnawing and ripping, what your sixth form biology teacher called mastication, gets the food down to size and increases the surface area so that the rest of your digestive tract can have its way with it. And when you eat something delicious and your mouth waters, that saliva is full of salivary amylase, the enzyme that begins to break down carbohydrates as you chew. Appreciate that saliva, because it's what makes food taste good: moisture helps your taste buds get a full flavour hit. (And even though this is the digestion chapter, let's be honest: taste matters.)

Once you swallow your food, you send it sliding down the rabbit hole of your oesophagus towards your stomach. We often reference the stomach when we talk about eating, so you may be surprised to learn that only minimal digestion occurs in your stomach. The job of the stomach is not to digest, but to mix and churn the food with gastric juices full of enzymes that begin to break down proteins and fats before the mixture is passed along to the small intestine, where most digestion and absorption takes place.

The stomach is a marvellous organ, but it can process only so much at a time. That is why if one eats, say, a giant cheese platter (this is purely hypothetical, of course), a wicked stomachache may ensue. Because the stomach does not empty all at once, but slowly, when you eat reasonable portion sizes, you help it do its job effectively (and save yourself from having to lie on the couch with a bloated belly for hours).

Once the food reaches the small intestine, the real work happens. Micronutrients like vitamins and minerals are already small enough that they don't have to be broken down; the small intestine can just absorb them as they are so they can begin to do their good work. Macronutrients take a bit more effort. In the small intestine, the pancreas adds its pancreatic enzymes so that the chewed, saliva-ed, gastric-juiced food particles can be further broken down into even smaller particles. These enzymes are made of proteins. The villi, tiny little fingers that jut out of the small intestine's surface, secrete more enzymes that help with the final stages of digestion. And even more help comes from the millions of tiny bacteria that live in your digestive tract and help your body get all the nutrition out of the food that you eat (we'll examine this more closely in Chapter 13).

Ultimately, after a brief stay in the small intestine:

- That baguette, a carbohydrate, is turned from larger sugars to single-sugar units like glucose, fructose and galactose. Those sugars will eventually be used for energy or stored as fat.
- That chicken breast, a protein, is broken down into its smallest unit – amino acids – and small peptides. Those amino acids will be used to build strong muscles.
- That pat of butter, a fat, releases fatty-acid chains, which will be used for energy and absorption of vitamins and minerals.

In order to be used by the body, your single sugars, amino acids and fatty acids, those treasures of digestion, must be absorbed into the intestinal cells. There they will continue their route on the digestion train. The next stop is the clearinghouse, the liver. The liver lives just above your stomach, and it plays a lot of important roles within your body. Guided by the insulin that your pancreas releases in response to the amount of glucose in your blood, the liver steps up to store glycogen (glucose that is stored long-term) for later use. It also breaks down toxic substances and stores a year's worth of vitamin A and vitamin B_{12}, a month's worth of vitamin D, and some vitamin K, iron and copper.

In addition, the liver receives deliveries of blood rich with nutrients that have just been absorbed through the intestinal walls. Those nutrients will be processed by the liver: fatty acids will be digested, sugars will be turned into glucose and amino acids will be separated from ammonia (which turns into urea and is eliminated from the body via your pee).

In the final stage of this energy cycle, your blood delivers the nutrients, along with oxygen, to your cells. Your mitochondria – the energy power-houses of your cells – break down the nutrients to provide the cells with a raw energy source. That process requires oxygen.

If you've ever built a fire, you know that it's not possible to get it burning without oxygen. Air rich with oxygen helps fires burn, because fire is the product of energy. Wood is burnt and energy is released into the air, heat energy that can burn your hand or boil water. Your body creates energy in a similar way.

When glycogen is burnt in the cells with the help of oxygen, it is broken down into carbon dioxide and water, and in the process, something called adenosine triphosphate, or ATP, is made. The creation of ATP is the ultimate

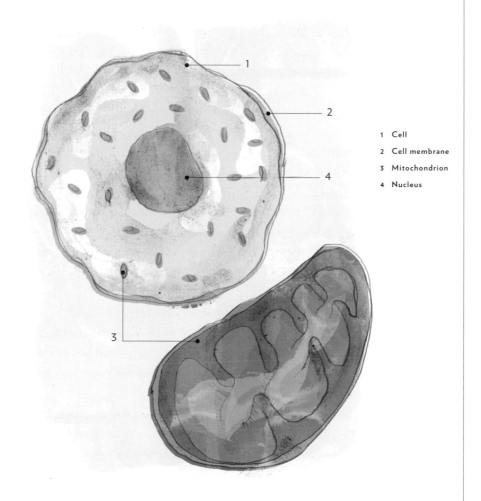

1 Cell
2 Cell membrane
3 Mitochondrion
4 Nucleus

goal of the entire food/digestion/energy cycle: like the heat energy produced by a burning fire, ATP is the energy that fuels cellular processes. It is often referred to as the energy currency of life, because ATP provides the energy for just about everything we do. Every time you laugh, every time your heart beats, every thought that you have, every movement of every muscle, the replication of your DNA – all of this is fuelled by ATP. And the by-products of cellular respiration, the carbon dioxide and water? Well, carbon dioxide is released every time you exhale. And you eliminate water and other waste products from your body in three ways. . . .

SPP: SWEAT, PEE AND POOP

The digestion process creates waste that must be eliminated. There are three main ways that your body eliminates toxins and waste: sweat, pee and poop. This is why it's so important that you drink plenty of water, eat fibre-rich foods and work up a sweat at least once a day. Because you are not only fuelling your body, you are helping it detox from the last fuel you gave it.

SWEATING IS COOL

More than cool, sweat is cooling. The main purpose of sweating is to regulate your body temperature. After all, your body produces heat – so you have to get rid of the excess. Without sweat, your body has a hard time cooling itself down.

You have two kinds of sweat glands: eccrine glands and apocrine glands:

- **Glistening Arms and Legs and Torso:** *Eccrine Glands*
  Your eccrine glands are all over your body. When your arms or sports bra are soaked after a spinning class, that's your eccrine glands. Those droplets of moisture are mostly water, with some salt and potassium, along with ammonia and urea – the substances released when you digest protein – and uric acid, which is a by-product of eating foods like anchovies, dried beans and dried peas.

- **Stinky Pits: *Apocrine Glands***

    Smell that? The beautiful, glistening sweat on your biceps and thighs isn't what makes you stink after a tough workout. Go ahead – give your sweaty arm a sniff after you lift. Nothing. The sweat that smells bad comes from your apocrine glands, which are located in your armpits and your groin – and that sweat is full of fatty proteins that the bacteria who live in those areas love to metabolize. So when you smell like a gym sock, it's not because your body is releasing terrible toxins; it's a natural process that's the result of the by-products of tiny little bacteria feeding on your sweat. Washing with a good antibacterial soap will help to combat the critters.

EL EM EN OH PEE

Going to pee, taking a tinkle, visiting the cloakroom, finding the facilities . . . whether you've been waiting in a queue for the loo for twenty minutes or getting free and natural in the local lake, a healthy and functioning urinary system is something to appreciate. And if you've ever had a urinary tract infection (UTI), I'm sure you know what I'm talking about.

The *urinary system* consists of your kidneys, your bladder and the ureters, the tubes connecting the two that are used to eliminate the urine from the body.

The *kidneys* are two bean-shaped organs that sit just above your waist on each side of your spinal column. As part of the urinary system, they filter our blood, retaining proteins, red blood cells and other things we need, and eliminating waste products and toxins in our urine. Basically, your kidneys make your urine. The *ureters* are narrow tubes that leave the kidneys and connect to the bladder. Your *bladder* is the organ that holds on to your urine until it is expelled from your body via the *urethra*, the tube through which your urine exits.

As I mentioned on page 88, paying attention to your pee – both how much and what colour – is an important indicator of your hydration level. If the clock strikes noon and you haven't hit the ladies' since your morning coffee, be sure to down a few glasses of water and eat plenty of water-rich fruits and vegetables at lunch.

CHECK THAT SHIT OUT

Lets talk about poop, baby! Yes, you heard me right. Listen, it's a natural and necessary part of being human. Now I know that some people don't like talking about their bowel movements, but here's why I'm talking about it: because your poop is a key indicator of your overall health. So let's get past the ick quotient and the modesty factor and become more familiar with what comes out the bottom of your beautiful bottom. Because poop is not 'eeeew'. It's knowledge.

Poop is our body's waste; it's what's left of the food we have ingested after our body has absorbed its energy and nutrients. Along with the waste from your food, it also contains waste products from your digestive juices and enzymes, as well as any potential contaminants that can cause illness or disease.

Everyone is different when it comes to how long it takes your body to digest your food and how long it takes you to eliminate waste from your body. Some people consider it normal to go three times a day, while others only go every other day. But what is most important is for you to know what is normal *to you*.

Do you know your poop well enough to notice if it has changed shape either gradually or all of a sudden? If you don't, you've got to start peeking before you flush, because the information you get from your poop can save your life! Yes, it can be that important to you. It can indicate if you have a serious illness or disease or simply a run-of-the-mill stomach bug. If you're not in the habit of taking a peek before you flush, I'm sure that there will be an

Humans fart an average of twenty-two times a day. Yes, even you.

THE COLOUR TEST

The food you eat determines the size, colour, consistency and quantity of your poop. Want proof? Eat a helping of beetroot and then pay attention the next few times you have a seat on the throne. Beetroot turns bowel movements a dark pink or red tone, so it's easy to see how long it takes for food to pass through your system.

opportunity in the near future. Just take ten seconds to look and make a mental note of the size, shape, shade and smell before sending it away forever. The more you know about your poop, the more you know about your health. And that's always worth paying attention to.

Here are a couple of general guidelines:

LOOSE, WATERY POOP: If you usually go once a day and all of a sudden you're going four times a day and the stool is loose and watery, you will know right away that something isn't right. You might be fighting off a bacterial or viral infection. Loose, watery stool can be a sign of too much fibre or possible bacterial/viral infection. And diarrhoea can cause dehydration, so you want to make sure to drink plenty of water. Because diarrhoea usually lasts only a few days, a longer bout can be an indication of something more serious, like a parasite, which you would have known if you had gone to see the doctor after day three.

HARD, DRY PELLETS: If you are someone who goes once or twice a day normally and then all of a sudden you don't go for two days, that is a sign of constipation. That's your body letting you know that you should take a look at what you've been eating, how much water you've been drinking and how much physical activity you've been doing for those past few days. Maybe you haven't eaten enough fibre, drunk enough water or taken enough exercise.

If you have suffered from chronic constipation and you've never called the doctor because you're embarrassed, ladies, that is not OK. Make an appointment now. If constipation is a new issue, keep an eye on it and, if it persists for more than a couple of days, you should see your GP. An obstruction in your bowels may not be sexy first-date conversation, but it is something you want to find out about sooner rather than later – and poop (or lack thereof) can give you that important first clue.

If you are prone to constipation you may think that the only solution is to use laxatives, suppositories or colonics to relieve you, but these 'fixes' can make issues worse if your body becomes reliant on their help. That's a vicious cycle that you definitely don't need to get into. What you do need to do is take a serious look at your dietary fibre and water intake as well as how much physical activity you've been getting. Constipation can cause the ratio of good bacteria to harmful bacteria in your digestive tract to become imbalanced, creating an environment for disease to thrive. And worst of all, it just makes you feel like crap all the time.

No matter what's going on with your digestive system at the moment, drinking plenty of water is always a good idea. It can keep diarrhoea from dehydrating you and help relieve constipation by keeping your poop moist and moving it along the digestive tract more easily. Once it gets to the colon, the colon will absorb the excess water to hydrate you and help the faeces come out of your body in the, ah, most comfortable way.

One of my tricks, as I've said, is that litre of water I put by my bathroom sink before I go to sleep. The next morning, after I brush my teeth, I drink the entire litre as quickly as I can. Let me tell you: it wakes up the digestive tract pretty darn fast. If you drink a tall glass of water in the morning and then you start to move and breathe, it acts like a wake-up call to the colon to start moving out the matter it's been storing in the rectum.

And that, dear reader, is the shit.

CHAPTER 13

SAY HELLO TO YOUR LITTLE FRIENDS

———

HOW DOES IT MAKE you feel when I say the word *bacteria*? If you're the type who carries a bottle of hand sanitizer in your purse and keeps an extra in the car just in case, you might already feel creepy-crawly. If you're the 'five-second rule' type who will eat popcorn off the kitchen floor, you probably didn't cringe at all. Either way, the fact is that planet Earth, the natural habitat of humans, is also the natural habitat of bacteria.

The ground is full of bacteria. The water is full of bacteria. The air is full of bacteria. Do you know what else is full of bacteria? You.

UH, WHAT??

It's true. Different species of bacteria are residents of every corner of this planet, from the banister on the staircase you used this morning to the deepest, hottest, most inhospitable places near volcanic vents at the bottom of the ocean to the spot just behind your left ear. Every layer of our world is full of bacteria, including your skin and your digestive system, and no amount of hand sanitizer will get rid of them. And that's OK. In fact, it's more than OK, because the bacteria in our world help support life as we know it.

Check it out: a billion or so years ago, the air on this planet was a poisonous mix of nitrogen and carbon dioxide. Nitrogen is harmless, but if we breathe in too much carbon dioxide it can kill us. Lucky for us, a group of bacteria called cyanobacteria helped shift the balance of the air on earth from deadly fumes to the oxygen mix that we breathe today. Without the cyanobacteria, this planet would never have been able to support our lives in the first place. And cyanobacteria still help us today by turning the carbon dioxide that we exhale back into the oxygen that we'll need to breathe on the next inhale.

Bacteria are our friends.

WELCOME TO YOUR MICROBIOME

Your body is made of trillions of human cells and is also home to trillions of bacteria – little single-cell, living organisms. The bacteria that live in your body are sometimes the kind that give you a stomachache, but more important, there are other kinds that lend support to your immune system, your digestive system and your cardiovascular system. Your body is basically a small civilization of bacteria: more than one hundred trillion of them. In fact, you have ten times more bacterial cells in your body than human cells. Ten times! Your nose is a neighbourhood for at least three types of bacteria. Your ears have their own bacteria. So does your digestive system – up to a thousand different kinds of bacteria can make themselves at home there. This is called your *microbiome.*

The bacteria that make up your microbiome are alive, just like you, but unlike you, they have only one cell each. They are so tiny that you could line up a thousand of them in a row and they would still fit on a pencil eraser. And there are enough of them in your body that if you could gather them all up and put them together on a scale, they would weigh around one and a half kilograms, roughly the same weight as a toy Chihuahua. Before you let your yuck factor kick in, pause and think about this: that colonization is part of why your skin doesn't crack, why your immune system can fight off other types of bacteria, why your body can digest that sandwich you had for lunch and why it can extract the nutrition from the rocket salad you had alongside it.

And here's something even cooler: nobody's bacterial makeup is exactly the same. Your microbiome, your own personal composition of microbiota, is uniquely yours.

HOW YOU GOT YOUR PERSONALISED BACTERIA

Your digestive system is full of bacteria, which are also called your intestinal flora. And as we've discussed, that's a good thing. Your intestinal flora have been with you for a long time, ever since you first made an entrance into this world.

But when you were a tiny thing growing in your mother's belly, you had no bacteria in your body. You didn't need any, because you weren't doing any digesting – all of your nutrition came through your mother.

If you were born vaginally, I learnt from Dr Maria Gloria Dominguez-Bello, a scientist and researcher at NYU, your first digestive bacteria came directly from your mum. If you were born by C-section, your first microbes came from the people who held you and the environment that you were in. Your microbial development was also affected by whether you were breast-fed or bottle-fed; the two methods of feeding influence microbial development in different ways.

As you grew and ate more foods and encountered more people and more places, your personal microbiome grew and changed with your life and the way you lived it. By the time you were two and a half, the microbes in your digestive system probably looked a lot like those of an adult.

Your current microbiome results from a number of factors: the bacteria you got from your mother, the environment you live in now and how much stress you are under. It is also influenced by the food you eat, and whether that food is whole food or filled with preservatives, or whether that food contains contaminants like antibiotics or growth hormones from cows or other conventionally raised livestock.

Here's the thing you want to consider: your microbiome can multiply, but it can also become depleted. That's what we're going to focus on here, because the health of the bacteria in your digestive system is directly related to your overall health. Research has proven that your digestive health is closely linked to your immune health: in fact, your gut has more of an influence on your immune system than any other part of your body. A healthy gut is your *main* weapon against any disease-carrying organisms that get into your body. That's why the health of your digestive system, which includes your microbiome, is so important.

THE PROBLEM WITH ANTIBIOTICS

Some bacteria – like the bacteria that cause strep throat – can make you ill. This you probably know. But here's something you might not know: Those same bacteria can be present in your body even when you aren't unwell. The difference between whether bacteria make you ill or not isn't their *presence*, but a bunch of other factors, like how healthy you are, how strong your immune system is and how strong that particular strain of bacteria is. Sound like a war? It kind of is – a battle between foreign, potentially dangerous bacteria and your resident protector bacteria as part of your overall immune response.

When your body is overtaken by bacteria that are making you ill, doctors will frequently recommend a course of antibiotics. Have you ever had an ear infection or other illness that required you to take a course of antibiotics? Antibiotic means 'against life' in Greek, and antibiotics are designed to kill living organisms – bacteria – that are making you ill. But our antibiotics aren't sophisticated enough to go on a search-and-destroy mission for only the bad guys. They kill any bacteria in their path, good or bad.

Because the good bacteria in your microbiome have been wiped out and your gut is no longer armed with enough of its very important colonizers, a course of antibiotics often comes with some unpleasant side-effects, such as poor digestion, diarrhoea or vaginal thrush, a type of yeast infection.

I had a conversation with Dr Martin Blaser, director of NYU's Human Microbiome Programme, who explained a lot to me about my microbiome, how important it is to my health and how science is working to uncover even more mysteries of the microbiome. I learnt that even if I haven't taken antibiotics in

COWS ON PENICILLIN

For the past few decades, many livestock producers have been giving low-grade doses of antibiotics to their cattle, because it prevents them falling ill, but it also has the curious side-effect of helping them to grow and gain weight faster. Now, scientists are researching why that happens – and how antibiotics and human obesity are related.

years, if I eat a lot of red meat I can still be exposed to the harmful side effects of antibiotics from cows that have been injected with them! One more reason for me to choose grass-fed, humanely raised meat.

The point is that antibiotics can be helpful – sometimes you really, really, need them – but you shouldn't get into the habit of asking your GP for a prescription every time you sneeze. When your healthy bacteria are wiped out, you're at an elevated risk for developing other illnesses, plus it makes it more challenging for your immune system to defend you and for your digestive bacteria to absorb the nutrients that keep you healthy.

THE PROBLEM WITH PRESERVATIVES

Do you know what else kills your healthy bacteria? Processed foods that are stuffed full of preservatives. The purpose of preservatives is to kill the bacteria that make food go off. But when those preservatives enter your digestive system, they also kill the healthy bacteria in our guts that help us stay alive. Eating whole, fresh foods makes preservatives unnecessary!

The accumulation of all of these things – antibiotics that you've taken for illness over your lifetime, processed foods stuffed with preservatives, meat treated with antibiotics – contributes to depleting the good bacteria in your digestive system, which by now you know are your best little friends. A dose of antibiotics taken for ten days has a big impact on your internal environment. But the slow drip of less obvious antibacterial measures, from commercially raised meat to preservative-packed biscuits, also has a significant impact.

EATING BACTERIA FOR HEALTH

If our microbiome can be depleted by *anti*biotics, it can also potentially be replenished by *PRO*biotics. When we eat healthy bacteria in the form of probiotics, we can help promote the balance of good bacteria in our gut. In addition to offering immune-boosting and disease-fighting help, some strains of bacteria in probiotics, like *Bifidobacterium infantis*, have been shown to be beneficial for digestive issues like irritable bowel syndrome (IBS).

Naturally fermented foods are a source of probiotics that humans have been relying on for thousands of years. Yoghurt, a great source of probiotics, has been eaten for six thousand years (those Mesopotamians knew all about good intestinal health!). Sauerkraut, which is made from fermented cabbage, is mentioned in ancient Roman texts. Kimchi, spicy fermented cabbage, has long been a staple in Korean diets. In fact, kimchi is so central to a traditional Korean diet that many Korean homes have kimchi pots buried in the backyard for fermenting homemade kimchi. Umeboshi plums are Japanese pickled plums that are eaten with meals for a salty, tart taste and some delicious digestive help.

About a hundred years ago, science started catching on to the inherent wisdom of these traditions. Are you noticing a trend here? Humans survived until now by eating whole foods rich in nutrients and other great stuff, like bacteria. No one had a microscope to know *why* back then; they just knew that eating what they ate helped them survive. It's only nowadays, when we have moved away from these natural ways of living well, that we struggle to understand their value and how to fit them back into our meal plans.

In the early 1900s, when Americans were just beginning their romance with processed, mass-marketed convenience foods, a Russian scientist named Elie Metchnikoff was studying intestinal bacteria. Metchnikoff, who lived in Paris, believed that bacteria were the fountain of youth. At the time, there were two opinions about intestinal bacteria: some believed they were a key part of digestion, while others believed they were harmful. Metchnikoff noticed that people in the Balkans, who ate loads of yoghurt, tended to live until well into their eighties, so he started eating yoghurt himself. Soon enough his friends started eating yoghurt, and then their friends, and eventually, eating yoghurt became a health trend in Paris. Some doctors even prescribed it to help cure a variety of ailments. Lucky guy—not only did his health improve, in 1908, he won a Nobel Prize.

Today the World Health Organization defines probiotics as 'live microorganisms which when administered in adequate amounts confer a health benefit to the host'. Scientists are queuing up to discover just how beneficial these bacteria can be and how much of them is needed for optimal health. (Eventually they hope to figure out how all bacteria – the bacteria in food, in our digestive systems, and on our skin – might be harnessed to create treatment programmes for specific health problems.)

When it comes to eating bacteria for health, probiotics have become a big business. They're not just sold in little vials in health food stores anymore;

they're in every store, with shelves in pharmacies and grocery stores stocking a range of foods, drinks and pills whose labels promise huge health benefits if you have just a bite or a sip of their fermented products.

But before you get too excited about the benefits of bacteria-enhanced products, be sure to check out their labels. Because the science around probiotics is so new, there's not yet a consensus on which strains are the most effective – so you've gotta do your research. Read your yoghurt label. *Lactobacillus delbrueckii bulgaricus* and *Streptococcus thermophilus* are bacteria that are great for making yoghurt, but your stomach's high acid content and bile break them down and keep them from having any kind of probiotic halo effect. A better choice is to look for *Lactobacillus acidophilus* and *Bifidobacterium*, because these guys can survive the intense conditions of your stomach for long enough to do some good.

Personally, I choose a rice-based probiotic drink that delivers fifty billion active *L. acidophilus* and *L. casei* bacteria. When I take my probiotics regularly, I feel good, and I feel like I am helping my body help itself.

PART TWO

FITNESS

The Body Wants to Be Strong

CHAPTER 14

THE BODY WANTS TO BE STRONG

—

'VE GOT GREAT NEWS for you: your body wants to be as fit and strong and gorgeous as *you* want it to be. That's right. Its instinct is to be powerful and resilient. The whole purpose of its existence, at least as far as nature is concerned, is to keep you strong and capable so that you can create more strong and capable beings, because nature programmes us for survival. Think about it – when something isn't healthy or strong, nature thins the herd. The fittest, literally, survive.

It's lucky for us – unlike our ancestors – that when we get weak and haven't fed ourselves well enough to be able to run fast, there isn't a lion lurking around the corner. But there are still things out there that are just waiting for us to become weak. Like disease – all kinds of disease. There are viruses that want us as hosts, like those that cause flu or measles or Ebola, and harmful bacteria, like salmonella and those that cause strep throat, that would like to hitch a ride until they meet their next victim. And then there are the diseases that we have a hand in, like type 2 diabetes and heart disease, which are direct offshoots of our diets and lifestyles. Instead of intentionally building bodies that are strong enough to fend off viruses, bacteria, and chronic illness, we are mindlessly creating bodies that are weak, prone to illness and just waiting to short-circuit. Basically, when we're not careful about our nutrition and fitness, we become a bunch of walking time bombs.

If you lived in a world where you had to run from lions in order to survive, where you had to hunt for your food, where you had to drag heavy loads without the benefit of machines, where you had to build your own home and dig your own fire pits, you would be strong or you would die. But you don't live in a world that requires you to use brute force all day long. You live in a world where you can drive through the drive-thru, flop down in a chair to work all day, and spend your evening on the couch, in front of the TV, before you crawl off to your cosy bed. Even if you work hard all day, your day-to-day living can make your body soft. And that softness is a modern-day killer, the equivalent of the savanna-dweller's lion (actually, the lion was better, since it kept people moving).

Every single one of us has the instinct in our body to move, to be active. Resting after exertion and getting enough sleep are crucial activities if we want our bodies to heal and repair, but we're meant to rest eight hours a night, not twenty-four hours a day!

Every single one of us has the instinct in our body to move, to be active. Resting after exertion and getting enough sleep are crucial activities if we want our bodies to heal and repair, but we're meant to rest eight hours a night, not twenty-four hours a day!

If you are not in the habit of being active, you are at risk for a number of ailments that would probably not be an issue if you just **moved**. Moving your body on a daily basis, continually throughout the day, is your body's instinct because it is essential to its well-being.

SOME INCONVENIENT TRUTHS ABOUT CONVENIENCE

In our work and our play, human beings have developed the habit of moving a lot less than we once did. Over the last century, decade after decade, with every technological innovation, we have relieved ourselves of the burden of physical labour.

Take, for instance, an activity as simple as washing clothes. Only since the late 1950s, when the washing machine was invented, would such a task seem simple. Before that, you would have had to use a tub that held the water and you would have had to move the clothes around by hand and then do the rinse and spinning cycles by hand. And since the dryers back then didn't work very well, you'd still have to hang the wash up on the line. Before that system was invented, you would have to go to a stream and rub the clothing over a ridged washboard to scrub it clean, then rinse it in the icy stream, wring it out and hang it to dry. And before that you would have washed your clothes by rubbing them on rocks and using sand to scrub away stains before you rinsed them in a stream, wrung them out and hung them to dry. Are you catching my drift?

Back in the day, when nothing was convenient, people had to manually do the tasks that we accomplish today with machines. Have you ever tried to wash a pair of jeans by hand? Of course you haven't. You have a washing machine that does it for you. If we used the time freed up by machines to do things like run and climb and dance, they would be convenient indeed. Instead, we've taken away the day-to-day physical activities that required us to utilize our bodies in a naturally consistent way, only to replace those functional movements with sitting and more sitting.

. . . somewhere along the way, we started to believe that the less we had to use our bodies to labour, and the more leisure we had in our lives, the better our lives were.

When people still used washboards, walked instead of drove, took stairs instead of the elevator, and got water from the well instead of the tap, there was no need to go to the gym and work out, because they were continuously active. Those undertakings used the energy and nutrition that we put into our bodies, built the muscles that we needed to do those activities and kept strength in our bodies. But somewhere along the way, we started to believe that the less we had to use our bodies to labour, and the more leisure we had in our lives, the better our lives were. At home and at work, we are sitting more and moving less, and that is having a negative impact on our health.

ALL OUR NEW TOYS

Many of the conveniences of modern life are probably things you take very much for granted. Who can imagine a world without ATMs? It just goes to show how quickly things have changed over the past few decades. The way we are entertained has changed. The way we access information has changed. The way we communicate with other people has changed. And most important, the way we take care of our bodies and prepare our food has dramatically changed.

Of course it isn't surprising that our lifestyles changed as a result. When we had to move to survive, we moved. Now, a lot of our survival and pastimes depend on the click of a button. Press a button and the music comes on. The TV comes on. The laundry machine starts. The dishwasher starts. The garage door opens. But there's no button that substitutes for good, old-fashioned movement.

THE SOCIETY THAT LOVES TO SIT

A team of scientists from three universities investigated the long-term affects of our national slowdown. First they examined how physical activity had shifted on the job from 1960 to 2008. Then they looked at how taking care of a home had changed from 1965 to 2010, specifically, how much energy was spent at home doing household chores like cooking, washing up and doing the laundry and other housework.

Now, we've already talked about how much more elbow grease was required to do housework before the advent of machines. And at the office, ladies, had you sought a professional career in the sixties? Nearly half the jobs available to you at private companies would have required you to move your body regularly. Today less than 20 per cent of jobs require that kind of movement. We used to walk, carry, lift, build, make. Now we type, text, e-mail, messenger and call. Machines have taken the sweat out of our days. This shift – from occupations that make us move to those that make us sit –

according to the study, is directly related to the weight that Americans have gained over the past fifty years. And because work probably won't require us to become more active as new technologies develop, the study notes the importance of having active lives OUTSIDE of work.

The home study found that as the amount of time spent on household work declined, obesity rates rose in women. Think about it. How many of our grandmothers were overweight or had diabetes in their thirties and forties? They were busy, digging in the dirt to grow their own tomatoes, pushing that heavy Hoover around and spending hours cooking dinner for their families. That was hard work!

Not that I'm advocating that we get in a time machine or mop until our biceps burn. I like the convenience of modern life when it means I don't have to wash my jeans with a rock in the river. But it's important to realize that the modern value of convenience has permeated the way we approach almost everything in life. This misplaced belief has stripped away the daily efforts that were arduous but still served us in a healthy way. *We are abusing the very privileges we have given ourselves.*

Essentially, we are all part of a giant experiment, with planet Earth as a petri dish, and we're just starting to realize that our choices may not be taking us in the best direction.

WELCOME TO THE EXPERIMENT

Until the last decade or so, we weren't fully aware of the dangers of an inactive way of life. We didn't understand the long-term implications of all of that convenience. Now, we're starting to catch on, and it's taken a frightening rise in obesity and disease to open our eyes. Essentially, we are all part of a giant experiment, with planet Earth as a petri dish, and we're just starting to realize that our choices may not be taking us in the best direction. We are using our own genius and creativity to become softer and sicker instead of stronger and better. Lazier, not leaner. And that is not OK.

1950s:

Many homes are fully electric now; many families have refrigerators, washing machines, coffee-makers and hoovers. The world's first television remote control is invented.

1960s:

Why go to a concert when you can listen to music with brand-new audiocassettes? Why play football when computer games were just invented?

1970s:

Food prep gets easier with food processors. Movie watching gets easier with VCRs. Instead of going to a cinema, just hit play.

1980s:

Computers for everyone! IBM PC. Apple Macintosh. The first 3-D video game. High-definition television was invented.

1990s:

No need to ever go to a library again with the emergence of the World Wide Web. No need to go out to make new friends. And staying in to watch a movie gets even easier with DVDs and Web TV.

2000s:

Finally, everything we need without leaving the couch! Your iPod holds all your music. YouTube gives you access to videos. Robots hoover the floor for you. The Web makes it possible to order all your meals, weekly shop, chemist's products, and clothing, without having to walk across a room. Instead of visiting friends, you can keep up with their every move via social media.

Sure, at first it made sense to have everything that we need at our fingertips. It's kind of amazing to be able to change the channel, order your dinner, call your mum, text your best friend, all while recording the four other shows you could be watching. But just think about how your personal technology has changed over the past decade. The speed at which we are incorporating this technology into our lives is increasing exponentially. There are apps for more parts of our lives than we actually have time to live.

It truly is a miracle, and convenience offers many benefits. But our muscles and our bones are NOT benefiting from all of this convenience, and neither are our hearts or our lungs or our brains. Our bodies are built pretty much the same as they were when we had to rub two sticks together to make a fire. Nothing has changed except the world we live in! And that turns out to be a *huge ginormous problem.*

The habits we don't even realize are habits are making us ill, slowly and surely. If we want to thrive instead of just barely survive, we must embrace the value of *trying*, the value of *sweating*, the value of *moving*. Our bodies need exertion, movement, effort, all the things that we usually try to avoid. Convenience is not an acceptable foundational value for society. It's a disease.

The conclusion for us today is that any energy we save by being able to toss our jeans in the washer and dryer should be spent running and hiking, not wasted away sitting and watching.

We need exertion if we want to survive.

GET MORE EXERTION, EVERYWHERE, ALL THE TIME

Exertion – using your body, using your energy, using your muscles, aiming to break a sweat – is easy to find when you know where to look. You can exert yourself anywhere! I always look for the places in my day-to-day life where I can add a little more exertion, try a little harder, move a little faster. Here are a few ideas for adding more exertion to your day:

- **If you can, walk.** If you have to drive, park a little farther away. At the office, get up and walk to someone's office instead of sending an e-mail. Volunteer to go to pick up lunch. The faster you go, the more you exert yourself.

- **Take the stairs.** Better yet, run the stairs! Or go slowly and focus on each step, really pushing through your butt and leg, squeezing them tight as you climb. Because whenever you contract a muscle, that is exertion.
- **Multitask.** When you're standing in the kitchen waiting for your toast to pop up, why not do some standing push-ups off the counter? Here's how:

1. Face the counter.
2. Put both hands on the counter a little wider than shoulder width apart. Take two steps back or until your arms and hands are supporting the weight of your body and your feet are just behind you. Now bend your arms at your elbows, so that your chest comes closer to the counter, keeping your body straight by keeping your stomach tight and legs straight. Take your chest as close to the counter as you can and then push back up, straightening your elbows. Do that ten times or as many as you can until you feel like you can't hold your form.

- **Don't just sit there while you watch TV.** Simply getting up and down off the floor is exertion, so you can watch your favourite shows while you exert yourself by doing a combination of sitting and standing:

1. Sit down on the floor, stretch your legs out in front of you, reach for your toes, count to three, then stand up, reach over your head, count to three, then sit back down, and do it all over again.
2. Do this twenty times consecutively, making the standing and sitting actions as quickly as you can, to elevate your heart rate.

- **Good old-fashioned jumping jacks.** They're one of the best ways to exert yourself without needing anything other than the ground beneath you. We all know how to do jumping jacks; I'll spare you the details here. Try doing twenty to give your heart a little boost in the middle of the day.

As little as thirty minutes of cardio three to five days a week can add years to your life. So if exertion is not a regular part of your life, you need to add it to your schedule. Just try it! If you don't have thirty minutes in the morning, aim for fifteen, and then move again for fifteen minutes in the evening. Or break it up into three bouts of ten. According to fitness and nutrition expert Kathleen Woolf, a professor at NYU's Steinhardt School of Nutrition, the benefits start immediately, and they keep on coming. . . .

WITHIN SECONDS OF EXERTING YOURSELF ...

- Your heart rate increases.
- Blood is delivered to your muscles.
- You start burning carbs and fat for fuel.
- You get an almost immediate mood boost.
- You breathe faster and deeper, making more oxygen available to your working muscles.

AN HOUR LATER ...

- You've strengthened your immune system.
- Your mood is still boosted.
- Your body continues to burn energy at a higher rate (your metabolism is increased).

THAT EVENING ...

- Your muscles are recovering and rebuilding.
- Your blood lipid (cholesterol, triglycerides) profile will improve.
- Your body will clear glucose more rapidly from the blood, which prevents heart disease and diabetes.
- You feel alert and focused.
- Your sleep quality will improve.

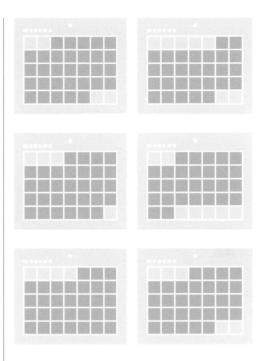

A WEEK LATER ...

- You have improved your endurance and aerobic fitness (you can go longer and harder than before).
- Your body is benefiting from a stronger immune system, a better mood and lower blood pressure.

THREE TO SIX MONTHS LATER ...

- You've improved the fitness of your heart and lungs.
- Your heart rate is lower at rest and recovers more quickly after exercise.
- You have improved the size and strength of your muscles.
- You've decreased your body fat.
- You've reduced your risk of diabetes, heart disease, cancer and osteoporosis.
- You've reduced your risk for depression, anxiety and stress.
- You've improved your overall quality of life.

When I move, when I am building the strength my body craves, I feel happy, clear, vibrant, alive, energized and powerful. I feel good. I feel like ME.

I recently found myself on a phone call with someone who was suggesting a plan that sounded terrible to me. This person was pushing hard. I was not enjoying the call; it was making me angry and stressed.

"So?" the person kept saying. "What do you want to do?"

I didn't want to say something I would regret, so I decided to hit PAUSE.

"Let me call you back," I said.

And I hung up that phone, put on my trainers, and got on the elliptical machine. I got a little sweaty. I gave my body what it needed: physical exertion instead of emotional and mental stressors.

And voilà! My stress began to evaporate. I was able to clear my mind, release my stress, sweat things out and get clear on what I wanted to do, because my body had moved and my endorphins, my happy hormones, had been released. After I exercised, I could breathe again. I wasn't stressed. And I was able to approach the situation completely differently, not just in what I said but in how I *felt*.

Had I not taken that movement break, I would have said things differently, and the outcome would not have been as beneficial as it turned out to be.

When you get moving, your heart races . . . you feel exhilarated . . . excited . . . like you're really *alive*. And when you're done, your body hurts in that amazing way that lets you know you really used it. Your brain is sharper. Your senses are more alert. That's the kind of stuff that makes you feel good about your day.

That is the power of exertion. It makes your body strong. It makes your heart strong. And it makes your mind and your resolve strong, TODAY, RIGHT NOW, every time you just get up and MOVE.

CHAPTER 15

THE MAGIC OF MOVEMENT

LIFE IS ABOUT MOVEMENT. Not just getting from point A to point B, but revelling in the journey. Think about a dancer leaping across a stage, a child running joyously across a field, a gymnast hurtling through the air. Movement is magical. When humans move, we become gazelles, like the dancer; cheetahs, like the running child; defying gravity for just a moment like the gymnast before we stick the landing. When we are in motion, we connect with our own natures, where thinking comes second and all you are conscious of is the pure sensation of being in your body.

Are you inspired by the grace and glory of the Olympics? The primal energy of a rugby match? Even just *watching* movement connects us to our bodies. But you don't need to be an Olympian to skate, tumble or jump. You don't need to be five years old to know what it feels like to run through a field joyously.

It is the responsibility of every human, not just dancers, sports stars, kids and Olympians – to move. To embrace movement. To become movement. To live movement. Moving can turn a regular Monday into an extraordinary start to your week. It can turn Tuesday into a quest, Wednesday into an adventure, Thursday into a triumph, Friday into a feat.

No matter how you spend your days, claiming your right to MOVE is like claiming your FREEDOM.

I did not like my body when I was a kid. I was all skin and bones, all arms and legs. I was really, really skinny, and the other kids let me have it because of that. I hated being skinny. Being made fun of feels terrible, no matter what the reason is, no matter who you are. I have a lot of girlfriends who struggle with their weight, and they all remember feeling self-conscious about being heavier than the other girls as young women. Skinny was their dream. I was on the other side of the spectrum: I wished that I had curves. Dreamy, lovely curves. Skinny or fat, when you're at the extreme, being made fun of leaves a psychological scar.

By the time I was in my twenties, I was so accustomed to being scrawny that I didn't give any thought at all to taking care of my body. I ate a lot and didn't gain any weight, so I didn't think I had to do any activities to build my physical fitness.

I was twenty-six years old. I had just quit smoking. I had poor eating habits. I had no strength.

Then I was cast in the *Charlie's Angels* movie. It was autumn, and Drew Barrymore and I arrived to train for our roles on set. (Lucy Liu hadn't come yet because she was finishing another movie.) Cheung-yan Yuen, who was our martial arts master, was there, along with all of our trainers. We were so excited. Woo hoo!

We had no idea what we were getting ourselves into.

Cheung-yan Yuen began to speak, and the interpreter translated his words for us.

"Today," he said, "I'm going to introduce you to your new best friend. You're going to learn to love your new best friend. You're going to have him with you all of the time. You're going to cherish him. And he's going to be a part of your life."

We were so excited. We looked at each other, as if to say, who could it be?

And Cheung-yan Yuen said, "Your new best friend is pain."

I looked at Drew, and she looked at me, and then we both looked at Cheung-yan Yuen, and then we looked at each other again, both of us thinking, *Did he just say pain?*

"Yes," Cheung-yan Yuen said. "Pain. You're going to learn to love pain."

Then he explained that he was going to put us through so much pain that we weren't going to be able to see straight. And he was not joking. The amount of physical work that we had to do for the next three months was so intense and so painful that it was also completely transformative, physically, emotionally and mentally. Thanks to my teacher, I was forced to learn just how capable my body could be. I was forced to push it way further than it had ever gone. I learnt that pain is temporary, but strength is lasting. I was forced to build the strength that my body had always craved.

The first week, I thought I was going to die. By the end, I felt like a superhero. Becoming strong made me feel so powerful, so capable, like I could do anything. And for the first time, I understood what it meant to be connected

Becoming strong made me feel so powerful, so capable, like I could do anything. And for the first time, I understood what it meant to be connected to my body.

to my body. That skinny frame, the body I had been ashamed of and wished away and wanted to trade in for a curvier model, was actually a strong, powerful body. And it was MY body.

And that experience – learning to connect to my body, to love my body, to truly live in my body – has been the foundation of everything I have done since. Everything. My career. My relationship with my family. My relationship with myself. I showed up every day, and I did it, no matter what. Even if I didn't want to do it, I still did it. And that built the discipline that I needed to do it then – and to know that I can do anything that I set my mind to now.

That's what I want for you: to understand that you can do anything if you set your mind to it, and that a healthy body, supported by eating well and moving a lot, is the vehicle that will get you there.

YOU ARE HOW YOU MOVE

When it comes to your nutrition, you are what you eat. When it comes to your fitness, you are how you move. Movement shapes your muscles, strengthens your heart and lungs, sharpens your brain and elevates your mood. When you've exerted yourself all day – I mean, worked really, really hard, nothing feels better than relaxing afterwards, when you've *earned* that rest. When you spend all day running around, and then you sink into a comfortable couch, you feel like you're on a cloud. But when you've done nothing but sit around all day, you feel terrible, don't you? It's like your body is made of lead.

I move because sitting around all day reminds me of being sick, and lying around for too long makes me feel like I actually *am* sick. Because being lazy ultimately doesn't feel luxurious or indulgent to me . . . it makes me feel foggy, listless, sad, depressed, bored, tired, insatiable.

Sometimes when you're sitting around, your body wants energy, and you misinterpret it as hunger and have a snack. But here's what your body wants: to MOVE. Your body gains energy from movement. Your body is saying, *I need to stay awake*. And what it really wants is oxygen. Because oxygen is a form of energy for the cells in your body, your muscles, ligaments, tendons, and organs. If you're not giving your body oxygen, you're starving it, just as when you don't give it the right nutrition.

If you regularly feel tired during the day, it may be because you aren't getting enough sleep, aren't getting the right nutrition, aren't drinking enough water or aren't moving enough. A good night's sleep should carry you through your day if you have good nutrition and good physical activity. Are you a yawner? If you're yawning in the middle of the day, ask yourself:

- How long have I been sitting still?
- Have been drinking enough water?
- Have I eaten nutritious foods?

The only time I yawn is when I haven't been moving my body or if I'm dehydrated. If I've had a good night's sleep, I know that a yawn is a sign that I need to move or need more nutrition or need to have a nice tall glass of water – or all of the above! Because, just as Isaac Newton, the father of physics, said, a body in motion stays in motion. When you live actively, you *feel* active. And that helps you to keep moving.

WAYS TO THINK ABOUT MOVEMENT

There are a few ways to think about how we move our bodies: training, being active and being sedentary. *Training* means scheduling systematic forms of exercise that build muscle and develop your cardiovascular system for an intended result. Being *active* means using your body all day long. Being *sedentary* means not using your body nearly enough.

If you went to the gym this morning, you're probably thinking to yourself: I'm active! Of course I'm active! I ran five kilometres before work! But even if you went for a run this morning, if you spend the rest of the day sitting in a chair at the office, you aren't as active as you may think.

WHAT DOES IT MEAN TO BE SEDENTARY?

Sitting on the couch is sedentary. Working at a desk is sedentary. Anytime you are sitting, sitting, sitting and not moving, not giving your muscles a chance to do their thing and not giving your heart a jolt you are being sedentary. And being sedentary comes with a lot of health risks, such as:

- **Back and muscle strain:** If you've ever spent all day in a car or at a desk, you know that sitting for hours on end can result in tight and stiff hamstrings (more on those muscles in a bit), and tight hamstrings force your back to work extra hard.
- **Heart disease:** Early research has found that people who work at jobs that require sitting, like mail sorters, have a higher incidence of cardiovascular disease than people with jobs that require standing and moving, like postal delivery workers. Even scarier, sitting time and sedentary behaviours have been associated with increased risk of death from cardiovascular disease.
- **Obesity and diabetes:** In a large study of female nurses, television watching was associated with an elevated risk of obesity and type 2 diabetes.

BREAK IT UP

We sit in the car, at our desks and on our couches – it's unavoidable. The key factors to think about are: for how often and for how long? Recently, researchers looked at how breaking up sitting time with short bouts of light- or moderate-intensity exercise affects health. Over a five-hour period, volunteers were asked to do one of three things:

- Just sit there;
- Sit, but take a two-minute light-intensity activity break every twenty minutes; or
- Sit, but take a two-minute moderate-intensity break every twenty minutes.

Well, guess what they found? Just a single day of uninterrupted time sitting can be hazardous to your health. At the end of the five hours, the people who had been sitting the whole time had higher blood glucose and insulin concentrations after eating compared with those who took the activity breaks. That means that just one long bout of sitting increased risk factors for diabetes.

So try to sit less. And anytime you're in a situation where you have to sit, break it up! Regular movement may be just as important to your health as increasing your physical activity. So get up and move! Take a break from the computer every twenty or thirty minutes. Get up and walk around while you're talking on the phone. Get off the couch during a TV commercial and climb up and down the stairs or dance around.

WHAT DOES IT MEAN TO BE ACTIVE?

Chopping vegetables or cooking dinner is active. Walking the dog around the neighbourhood is active. Walking around the mall to find the perfect pair of shoes is active. Standing instead of sitting on a bus or train is active. Make sure your activities of daily living are *active*. If you have a desk job, stand up and walk around whenever you can. As we learnt from the study in the box above, simply getting up and moving around for two minutes can dramatically improve your health and well-being. Being active should also include limiting nonwork screen time to less than two hours per day. If you spend all day working at a glowing computer screen, give your eyes and your body a break by being active instead of spending the rest of your evening typing and staring.

My personal way of being active is to just get up and do things whenever possible. If I'm on set, I jog over to the next trailer instead of sending a text. If I'm at the airport, I take the stairs instead of the escalator. Anything I can do to move!

WHAT DOES IT MEAN TO TRAIN?

When you exercise frequently in a way designed to make your muscles stronger and build cardiovascular endurance, you are training. Weight training, for example, builds muscle, increases strength and helps maintain bone density. Cardiovascular training gets your heart rate up and benefits your entire cardiovascular system, because when your heart is strong, it pumps more blood with each beat, delivering oxygen to tissues and removing waste products more efficiently. Cardiovascular training and strength training are also linked to decreased risk of chronic diseases.

When you are training, your body adapts to meet your energy needs. Because training requires a greater concentration of available energy than being sedentary, as your body adapts to your new routine, you become a fat-burning machine. This is amazing news for everybody who is training, whether your goal is to lose some weight or to complete a triathlon. But this result can only be achieved with a consistent routine.

If your body isn't accustomed to regular exercise when you first begin to train, the main fuel it burns is carbs. When those carbs have been used up and you run out of glucose, you become fatigued, and pretty soon you're done. But

EVERYBODY HAS TEN MINUTES

Recent research has demonstrated that about 150 minutes a week of moderate-intensity physical activity is associated with lower rates of heart disease and premature death. That's thirty minutes a day, five days a week – and those thirty minutes can be split up if you prefer. Research indicates that even short bouts of exercise, just ten minutes each, three times a day, benefit health and reduce risk for chronic diseases.

if you stick to a routine of training for at least three months and your body becomes a trained body, it learns to rely less on carbs. A trained body has better access to its fat stores. With what you know about carb storage (we don't have much) and fat storage (unlimited), this makes a lot of sense for bodies that need to keep going longer, because there is more energy available for use.

The training adaptations work deep down on the cellular level, where your mitochondria (cellular energy powerhouses) are. When you train and lose weight, you give your skeletal muscle more mitochondria and more mitochondrial enzymes. This adaptation makes you better at mobilizing fat from your body and getting it to the muscles where you burn it.

For athletes, accessing fat stores means that they can train longer without fatiguing, so they're able to gain endurance. Trained athletes actually begin to store more fat in their muscle, where it is easier for the body to access for fuel than belly fat or hip fat. When the fat is in your muscles, it's right there for the taking, like a healthy snack sitting in your fridge.

And for people who are training to lose weight, it's also great news: When the body starts to burn fat, you not only lose weight but also change your body composition – meaning you gain muscle and lose fat (which usually translates into inches lost). Whatever your goal – weight loss or athletic endurance – training helps your body burn fat so you can get where you want to go, whether it's across the common or up that mountain.

CONSISTENCY IS EVERYTHING

Unfortunately, this incredible process also works in reverse. That's the challenge for us humans: to work to achieve something and then stay consistently at that level. Because as soon as you stop training, your body knows, and it reverses the process. You make fewer enzymes. You have fewer mitochondria. You burn fat less readily.

If you don't keep up with your training, your hard work disappears. And if you want to be able to keep up with your training, you must take good care of yourself, pushing yourself as much as you need to improve, but pulling back before you injure yourself.

Being consistent doesn't mean that you are always training in exactly the same way, with precisely the same amount of intensity. Life changes, our bodies change, our circumstances change. I know that for myself; over the last fifteen years, I've cycled through different levels of fitness.

Sometimes I'll train intensely with a specific goal in mind. If I'm going to go on a snowboarding trip, I might want to get my overall body in good strength, especially my quads (thigh muscles) and my glutes (butt muscles). When I'm crouched on a snowboard, hurtling down the side of a mountain at top speed, I want good endurance in my muscles so that I don't fatigue quickly. And my core needs to be strong for manoeuvring. And I need to feel strong all over my body in case I take a bad spill, because strong muscles help protect bones from injuries.

Most of the time, I'm focused on maintaining a level of fitness that fits in with my everyday life. There's a lot going on, and I need to be capable of doing everything from lugging my baggage around the airport (a great way to get muscles engaged before long, sedentary flights) to carrying grocery bags or cords of wood or anything else that may have to be moved, carried, hauled, pushed or pulled. And how about all the energy you need to get some solid playing in? I love to play with my nieces and nephew, and my fitness needs to be up to par in order to keep up with them. And even cooking for friends, because spending hours chopping and sautéing needs standing stamina and slicing and dicing power. I also love doing my chores. Hoovering and sweeping don't just have to benefit your floors! These household chores are great helpers for your heart and muscles. The point is to try to find activities that must be done in everyday life and make the most of them by using them to also benefit your overall physical well-being.

Of course, the kind of muscle I need to wash the floors and counters is not the same as the level of fitness I need to develop when I'm training for a film that will require challenging stunts. That requires a little heavier lifting!

The point is that life changes, and we need different types of strength at different times. But the one thing that must remain consistent is MOVE-MENT. In my own life, although it's not always easy or possible for me to keep to a routine, I've made the commitment to myself to keep training and physical activity a part of what I do on a daily basis regardless of what else is scheduled on my calendar. Just as I brush my teeth every day no matter what country or city I may find myself in, I always find a way to move.

The only time to stop moving is when your body is healing and repairing, because certain injuries require rest. But usually, generally, movement is

ALWAYS a yes. You must never just let it go. And if you keep it up for years, you'll see how the consistency pays off.

Because I've had years of consistency, if there is a period of time when all I can do is twenty minutes a day every other day, it's okay, because I know as soon as my schedule allows it I'll be back to an hour a day with a more focused attention on training. And because I am regularly doing things like hauling luggage and the shopping!

Make movement a part of you.

Make a commitment to yourself that you will always give yourself physical activity or focused training, no matter what. Once you do that you will realize that it's something that you can't live without. Nothing is truer than that, your body requires it and so do your mind, your heart and your happiness.

LET'S REMEMBER WHAT IT FEELS LIKE TO PLAY

When I was a kid, I loved running. I loved playing softball. We were always out in the street playing with all the other kids, night and day. We just were *active*. We didn't think about how many calories we were burning or wonder whether softball would make our bodies more or less toned than riding our bikes. We just ran and swam and sweated, joyously and often, whenever we weren't at school or in the house doing chores. Our baby arms and legs grew longer and stronger and our muscles were stringy from use. But like a lot of kids who are active in their playtime, as we became adults, that kind of activity slowed and then disappeared. Play got left behind. And if the only movement you have is related to play, and there's no more playing – well, there's no more moving.

When we are adults, our playtime isn't scheduled for us, and unless you have a job that requires a *lot* of physical vigour, like moving boxes or building houses or running around after children who *do* know what it means to play, it's so easy to watch your playtime – and your activity – disappear. When we think about fitness, we don't think about playing. We think about lifting weights at the gym, running on a treadmill and hiring a personal trainer. We think about calories burned, sweat sweated, time spent. But that kind of movement is only part of the equation. If you want to be a healthy person, you must move ALL OF THE TIME. You must remember what it feels like to move for the JOY OF IT. To move because you *can* move. And because unless you remember how to move, not just once a day for forty-five minutes or three times a week for thirty minutes, not just on Sundays, but always and often as opposed to sometimes and never – unless you remember to move, as you get older, movement is going to become a lot less comfortable and free and agile and a lot more challenging.

When we are children, our bodies are elastic. Unless we keep them elastic through frequent movement, our muscles shorten and tighten and shrink, following the patterns of our lives, whether we carry a heavy bag on one shoulder all of the time or walk around in heels more often than we should. Moving your body keeps your muscles supple, increases strength and flexibility and is part of how your body uses the fuel you give it when you eat good food.

The next time you groan at the idea of hitting the gym, think about what you did as a kid to move your body in a way that you loved. Instead of dreading activities you find dull, start *enjoying* activities that thrill you. Did you roller-skate? Go to a rink with your friends, or borrow a pair of rollerblades and head to the park. Were you a daredevil on your bike? Get into the garage, find that dusty bike and hit the road. Think back to what life was like when you knew instinctively that playing was the best possible use of your abundant energy, and get out there and move.

CHAPTER 16

ENERGY IN, ENERGY OUT

———

THE ENERGY YOU SPEND living, breathing and moving all comes from the carbohydrates, proteins and fats in the food that you eat. That's the energy that powers your movement, your thought, your cellular repair and everything else that happens in your body.

These days we've become hyperaware of the balance of energy we take in versus the energy we expend – in other words, how much food we eat and how much (or how little) we move. For most of human history, this was a homeostatic, or self-regulating, process. We stayed pretty much within a set-point range of weight, with help from our hormones, which tell us when we're hungry and full and tell our body how and when and where to store fat so we can use it later. Today our self-regulating process has become totally screwed up by the rapid increase in the availability of low-quality calories, like refined grains and added sugars, and the worst offenders, sugary fizzy drinks and sports drinks. And while I'm not a big fan of calorie counting (in or out), we do need to be mindful of this balance, because when we blast our bodies with huge amounts of processed foods and sit for ten hours at a time, our bodies lose the ability to regulate our weight within a healthy set range.

The balance between our nutrition and physical activity is basically a mathematical equation. The energy you put into your body in the form of calories is added, and the energy you use in the form of physical activity is subtracted. Weight gain is the result of ingesting more fuel than your body needs immediately, so it stores some of the extra fuel as fat for later use.

Weight loss is simply the result of your body using up more energy than you have consumed, which means that your energy stores are empty and will need to be replenished.

Ideally, if you eat well and you live an active life, energy comes in and energy goes out in proportion to each other, and your weight remains at about the same place.

YOUR HORMONAL BODY

Hormones are related to your moods, your sleeping patterns, your sexuality and your appetite . . . not to mention your metabolism, your weight and where fat is stored on your body. Some of the key chemical messengers that help determine body weight and composition are:

GHRELIN: THE APPETITE-STIMULATION HORMONE

Ghrelin, made by the stomach, is the hormone that stimulates appetite. Blood concentrations of ghrelin are high right before a meal and then drop after a meal. When a body is losing weight, the stomach is triggered to make more ghrelin, in turn triggering the desire to eat, which can make it difficult to keep weight off. Obese individuals tend to have higher concentrations of this hungry-making hormone.

LEPTIN: THE APPETITE-REDUCTION HORMONE

Leptin, made by fat cells, makes its way to the brain with a message to reduce appetite and stimulate energy expenditure. The more fat on the body, the more leptin found circulating in the blood. In theory, this extra leptin should result in weight loss. Unfortunately, overweight and obese individuals aren't as sensitive to the effects of leptin. And further complicating the scenario—when a body loses weight, leptin concentrations fall, making it harder to maintain a lower body weight.

OESTROGEN: THE SEX HORMONE

Oestrogen is made by the ovaries, and plays a role in body-fat distribution. Oestrogen ensures that women of childbearing age store more fat in the lower body ('pear-shaped'). During the menopause, oestrogen concentrations drop and body-fat distribution changes, with more fat being stored in the abdomen area ('apple-shaped').

CORTISOL: THE STRESS HORMONE

Cortisol is produced by the adrenal gland in response to stress. Cortisol helps the body release fuels (e.g., glucose, amino acids and fatty acids) to counteract stressors (e.g., illness, injury). But emotional or mental stress typically doesn't require the use of extra fuel. High cortisol concentrations also increase abdominal fat, the type of fat more closely linked to diseases such as diabetes and heart disease.

There are three main ways that your body burns the energy you eat: rest, food and movement.

Have you ever heard anyone complain about their metabolism being too slow? Many people associate metabolism with how quickly they burn calories. But that's a huge oversimplification of how both metabolism and energy function in the body. Your metabolism isn't just about burning calories to lose weight. It's about using fuel to LIVE.

So instead of seeing food as the enemy that tricks your body into becoming fat, and exercise as the trick that gets your body to become thinner, let's focus on how our body's energy balance actually works.

REST: Your body burns energy even while you are resting; 60 to 70 per cent of your energy is burnt by your body's basic life processes. Even when you think you aren't doing anything, you're actually really, really busy just being alive, and that takes energy. The energy burn that happens when you aren't thinking about it is called your resting metabolic rate (or basal metabolic rate).

Your metabolism itself needs energy. Your resting metabolic rate supports all of the metabolic processes – the chemical reactions in your cells – that support your life. It takes energy to maintain body temperature, build new cells, keep your heart beating and your blood circulating, keep your lungs inhaling and exhaling and support everything else your body does automatically.

The rate at which your body burns energy while you are resting can be influenced by many factors. The more muscle you have, the more energy you burn when you are resting. Young people have a higher resting metabolic rate than older people. When a woman is pregnant or breast-feeding, her resting metabolic rate is also higher.

When you don't eat enough food, your resting metabolic rate slows down to conserve energy. That's why excessive dieting or simply not eating as often or as much as you should can make you fatigued and affect your ability to concentrate and think clearly – your body and your brain need energy. (Remember that story from my modelling days? I was a zombie version of myself when I was running on empty. Now I know it's because my brain literally couldn't function without food!)

FOOD: The energy that your body uses to digest, absorb, transport, process and store the food that you eat is called the thermic effect of food (TEF). The thermic effect of food typically represents about 5 to 10 per cent of your total daily energy intake. For example, if you eat two thousand calories over the course of a day, your body will spend one hundred or two hundred of those calories just to process that food. Isn't that cool? So eating smaller amounts of wholesome foods more frequently to fuel your body actually increases your metabolism and burns calories! So when you have a handful of almonds as your afternoon snack, you get an instant energy boost and some added calorie burn! That's why eating nutritious food when you are hungry – instead of starving all day and then eating all of your food in one humongous meal – is a more effective strategy for maintaining a healthy weight.

MOVEMENT: Here's the part where you can really get involved: get your butt moving. The amount of energy expended with physical activity is the variable that you are really in charge of. This component includes your activities of daily living, from showering to shopping for groceries, and your planned physical activity, like going to the gym or riding a bike. Physical activity typically represents 20 to 30 per cent of total daily energy output.

Calories are just a way to measure energy. You can measure any kind of energy with calories, even the energy of a motorcycle going at top speed.

BALANCING YOUR PERSONAL ENERGY EQUATION

If your body is already storing energy for later, whether it is five, ten or fifteen kilograms, then your personal energy equation is unbalanced. If you have no stores of energy and you're so skinny that your mother can count your ribs from across the street, then your personal energy equation is also unbalanced. Your energy intake and your energy output must be in harmony if you want to keep your machine running smoothly.

BALANCING ENERGY

energy in = energy out . . . weight is maintained
energy in > energy out . . . weight gain
energy in < energy out . . . weight loss

ENERGY IN: Identify where you may be consuming extra, empty calories. By avoiding foods that are proven to wreak havoc on your system and contribute to obesity, you can have a meaningful impact on your energy-balance equation without ever breaking out a calculator app:

- Drink water and unsweetened tea instead of fizzy drinks and juice.
- Eat nuts and whole fruits instead of sugary snacks.
- Eat whole foods instead of processed foods.

ENERGY OUT: Perhaps your schedule has changed and you find yourself with less available time for movement, and that has resulted in a weight gain. Or it's wintertime, and you find yourself craving hearty food but not getting out to use that extra energy. You need to move! And you might find that the more you move, the more you crave whole, real foods, because movement connects you to your body, and being connected to your body helps you understand your body's real needs. I know that after I go hiking or running, all my body wants is something healthy and nourishing to help it replenish its fuel so I can have some more fun and movement later. If you need to increase your energy output:

- Train and build muscle (to burn fat).
- Move more throughout the day.
- Sweat at least once a day.

SOME FACTS ABOUT FAT

The number-one fact that you need to know about fat is that the fat that you eat and the fat that your body stores are *not* the same thing. Not all of the fat that you eat gets stored as body fat; but refined carbs like sugar *are* stored as body fat. So please, do NOT fall into the false-logic trap of thinking that eating a handful of tasty nuts is going to translate into cellulite! It is more likely that the no-fat, refined-sugar-laden fizzy drink you've been gulping all day will later appear on your body as unwanted pounds, long after those almonds have made their way through your liver and into your mitochondria.

When I was trying to understand how the fat that we store in our bodies affects us, I read *Fat Chance* by Dr Robert Lustig, which really showed me how the numbers on a scale and the health of our bodies are less related than the diet industry wants us to think.

I learnt that when you stand on a scale, the number you see is based on the total weight of your bones, your muscles, your subcutaneous fat and your visceral fat.

- **Bones:** the more your bones weigh, the better off you are, because strong bones translate into longer, healthier lifetimes.
- **Muscles:** the more your muscles weigh and the bigger and stronger they are, the healthier you are.
- **Subcutaneous fat:** this is the fat that lives on your butt, your thighs, your hips and elsewhere and lends you your gorgeous curves and provides your body with lots and lots of back-up energy. Subcutaneous fat makes up 80 per cent of your body's fat stores, and it is not the kind of fat that contributes to illness.
- **Visceral fat:** this is the fat in your abdomen that's packed in around your organs, like your liver, stomach and kidneys. Visceral fat also lives in your muscles. It makes you more predisposed to life-threatening illnesses and potentially takes years off your life. This fat also has negative effects on your brain, your mood, and your health.

The more you work out, the more you move, the more you strengthen your bones. The more you build your muscles, the more energy you burn even at rest, because more muscles = more fuel burnt. And that fuel burning will reduce your visceral-fat stores, so you burn the fat that can hurt you.

CONSISTENCY COUNTS

Do you know why it feels so good to move? Since your body doesn't store movement the way it stores energy, movement must be continuous and frequent. *It's all about consistency.* The benefits of consistent movement result in a happy, energized body.

There are many reasons why you need exercise and regular movement, and it's not just about fitting into your jeans. Your daily habits and movements have an impact on your skeletal structure and your muscles, and they affect how your body works at the cellular level. Every one of the trillions of cells in your body participates in chemical reactions that provide fuel, build bones and skin or other tissues and dispose of waste. You are the sum of all of these metabolic processes. An active, fit body has sharper thinking skills, quicker reflexes and a stronger immune system than one that is inactive. In Chapter 7 we looked at how insulin helps your cells convert glucose into energy. The more you move your body, the faster that conversion can happen. Physical activity plays a huge role in how quickly you can get that glucose in the cell. *Using your available energy helps your body access stored energy more quickly.*

When your cells are working efficiently, you feel AMAZING. When your metabolism operates efficiently, you feel POWERFUL. The key to all of this is eating for energy and then using that energy through MOVEMENT. The reason our body wants us to eat is because it wants us to move.

No matter how much you don't feel like getting off the couch and going to the gym, or training for your upcoming 5K, or walking home from work; no matter how sore your muscles are at the end of a workout, you always feel really good after you've exerted yourself. And those good feelings are related to how your body is working at the cellular level, because how we feel on the outside mirrors what is going on inside, and vice versa.

So: are you an active person or a sedentary person? As long as you remember that part of being active is *moving continuously through the day,* you can become an active person RIGHT NOW. Just move. Each moment is an opportunity, so take advantage of every single one. It's that simple.

MOVE YOUR BODY

Movement can be done all of the time, and it should be. Here are a few ideas:

- Do butt squeezes while you're brushing your teeth.
- Do lunges while you're waiting for the coffee to brew.
- Do calf raises while you're waiting for the train.
- Run up the stairs. Run down the stairs.
- Stretch your calves on the stairs.
- Walk to the next bus stop. Or the next.
- Do sit-ups while dinner is in the oven.
- Stretch during the commercial breaks.

CHAPTER 17

OXYGEN IS ENERGY

———

YOUR ENTIRE BODY, YOUR entire being, relies on the dance between breath and blood for survival. *Breathe in.* You've just sent a rush of air to your lungs, which will extract the oxygen that your cells need to survive, filling your blood with oxygen. *Breathe out.* You've just released waste, in the form of CO_2, clearing your blood so it can carry more oxygen through your body. Meanwhile, the oxygen-rich blood is being pumped from your heart and into your arteries, where it will be sent through your capillaries to your entire body, replenishing the cells of your brain, your liver, your fingers, and your toes.

As you have already learnt, oxygen is energy: it powers the cellular respiration that produces the ATP that fuels every action, movement and thought in your body. So when you feel yourself flagging, inhale a ready energy source, and keep that oxygen flowing.

Getting plenty of nice, fresh oxygen is something I think about a lot. Especially when I'm filming a movie.

When you go to the cinema, the film you watch when you settle back in those cushy seats may be set anywhere in the world, from New Zealand to London to Chicago to China. And it can look pretty glamorous and exciting after all of the scenes are cut together and projected onto the big screen. But here's a secret: it's a different experience on the set, while you're filming. Because whether the audience gets to ooh and ahh over mountains or jungles, deserts or bustling cities, most movie sets look alike. Movie sets in Paris look like movie sets in Hollywood look like movie sets in Indonesia. Truly. And what they look like are trailer parks. Huge, sprawling trailer parks, with trailers parked all over the place. And each trailer is an office where work gets done or a bungalow where someone lives. I live in a trailer, the other actors live in trailers, we get our hair and makeup done in trailers. And then there's the set, where we shoot the scenes, and the set is usually far away from the trailers, sometimes really far.

Usually there will be a golf cart that is available for me to get to and from the set, but depending on the location and the surrounding terrain, I will use the commute from my trailer to the set as an opportunity to move. I use it as an opportunity to walk briskly, or take a light jog, or sometimes even run. I don't like to just wander from place to place. I like to move with purpose. I look at every moment as a chance to pick up the pace of my movement throughout the day. I run to hair and makeup from my trailer. I dash back to my trailer if I've forgotten something. If I need to ask someone a question, I sprint to their trailer.

I move like that because I want as much oxygen in my body as possible. The more rapid the movement, the more oxygen gets inhaled, absorbed into my lungs and delivered into my cells through my blood. When I move and my heart pumps faster – when my heart rate goes up – my heart is pushing more blood around through my arteries and capillaries, full of more oxygen, full of more energy. And I know I'm really, really going to need that energy.

Shooting on set can sometimes mean an entire afternoon stuffed into a hot room that is full of too many people – which means not very much fresh air. Have you ever been in a warm, crowded, oxygen-starved room? That kind of situation can put you right to sleep. It's basically a recipe for naptime. But when I'm on set, I need to be alert, not sleepy. I'm not there to nap. I'm there to work. And that work requires me to be focused and aware, lively and excited. I must be present in my scenes with other actors. It takes energy to portray a character, deliver lines and engage with the material.

When I'm on set, I work twelve-hour days, minimum, and I need every drop of energy I can get. That's why, every time there is an opportunity for me to jump or dash or sprint, I do it. Because I know that every quick sprint from place to place, every stair I climb, every gulp of breath, is feeding my lungs and my blood and my heart and my brain, so that I can get through the day.

Moving makes my heart beat. It makes me breathe deeply. It wakes my mind up. And that makes my body feel good. Which makes me feel awake and alert, and it puts a smile on my face, because feeling energized makes us happy, and it takes only a short burst of movement to get there.

INHALE, EXHALE

When you take a deep breath, all of that life-giving oxygen is absorbed into your body through your lungs, two intricately designed sacs located on both sides of your heart, within your ribcage.

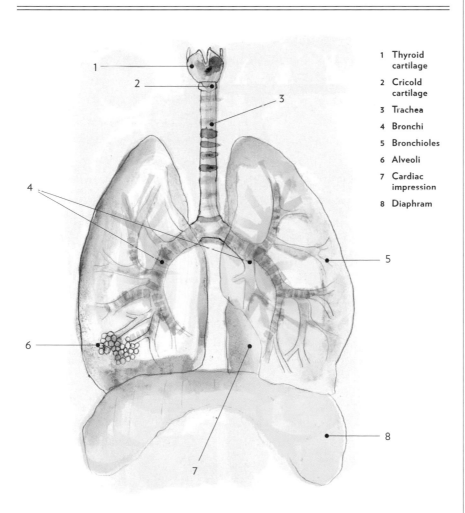

1 Thyroid cartilage

2 Cricold cartilage

3 Trachea

4 Bronchi

5 Bronchioles

6 Alveoli

7 Cardiac impression

8 Diaphram

Every time you take a breath, your lungs are assisted by your diaphragm and your ribcage—muscle and bone. The diaphragm is a broad, flat muscle just below your sternum and your ribs that helps control inhalation and exhalation, along with your ribcage, which moves with your lungs, expanding and contracting as you breathe.

As you inhale and fill your lungs with air, nitrogen, oxygen and carbon dioxide enter your mouth, move through your trachea, your windpipe, into the two branches of your bronchi, and then into your bronchioles, which keep dividing and getting smaller, like the roots of a tree.

Ultimately, the air winds up in your alveoli, which look like little bundles of grapes, and that's where the real work happens. The alveoli take the carbon dioxide *out* of your blood and put oxygen *in*, then send it throughout the body via two arteries that connect to the heart. This process happens again and again, throughout the day. And it speeds up when you are active. When you are running around, you breathe faster, because your body needs more oxygen to keep you going. Your breathing rate likewise slows down when you are resting.

Your lungs are made of a spongy, light material that can float in water; they are also very elastic, which is why they can expand without bursting when they are full of air. When you were born, they were a pink-white colour; as an adult, your lungs are closer to a dark grey.

Your two lungs are not exactly alike. They both have lobes, or sections, but the left lung has two lobes and the right has three; the right lung also weighs a little bit more than the left. Both lungs have an area called the cardiac impression, where the heart rests. The cardiac impression is a bit larger on the left lung, as the heart sits slightly towards the left within your chest. Now that's what I call good design.

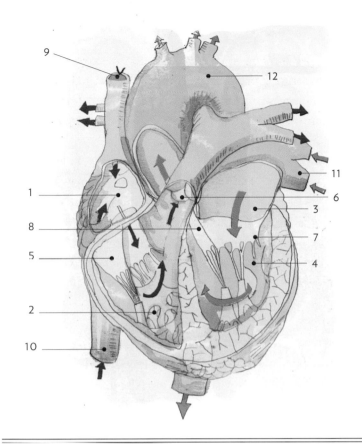

1 Right atrium
2 Right ventricle
3 Left atrium
4 Left ventricle
5 Tricuspid valve
6 Pulmonary valve
7 Mitral valve
8 Aortic valve
9 Superior vena cava
10 Inferior vena cava
11 Pulmonary veins
12 Aorta

YOUR BLOODY VALENTINE

The heart that beats in your chest does not look like a box of Valentine's Day chocolates. It is not adorable. But it is not meant to be, because it is not an accessory for Cupid. Your heart is a strong, gorgeous, internal organ with an incredibly efficient operating system.

That beautiful heart of yours is a muscle about the size of your fist, and it lives nestled between your lungs. The main job of your heart is to pump blood rich with oxygen from your lungs throughout your body, and it does its job more than a hundred thousand times a day.

Your heart is divided into four sections called chambers: the *right atrium, right ventricle, left atrium* and *left ventricle*. Your body's ingenious design includes valves to control the blood flow out of each chamber: the *tricuspid valve* for your right atrium; the *pulmonary valve* for your right ventricle; the *mitral valve* for your left atrium; and the *aortic valve* for your left ventricle.

When your heart beats, each chamber contracts, and its valve opens to let blood flow through. When the chamber is finished contracting, the valve closes, to prevent your blood flowing back in.

The right and left sides of your heart have different jobs. The right side pumps blood to the lungs, where it gets a fresh dose of oxygen and releases carbon dioxide. The left side takes that oxygen-rich blood and sends it on its way to your cells. Your blood enters the right side of the heart through two veins: the *superior vena cava* and the *inferior vena cava*. After blood is oxygenated in the lungs, it comes back to the heart via the *pulmonary veins*. The *aortic valve* releases blood into the *aorta*, which dispenses all of that life-giving blood to the rest of your body.

Over the course of your lifetime, in order to get all of this done, again and again and again, your heart will beat roughly *three hundred billion times*. The bright blood that rushes to the surface of your skin when you cut yourself shaving or skin your knee is just a drop of the five and a half litres of life-giving fluid that travels more than nineteen thousand kilometres every day through all of your arteries and veins and capillaries in order to give you the stuff of life, oxygen. Your blood also transports amino acids and hormones that build muscle and make you feel alert, hungry, horny or sleepy, plus all the nutrients that are supplied by your nutritional intake.

Cardiac muscle just pumps and pumps. The better you treat it, the better it will treat you. And here's the amazing thing about your heart cells: they beat *individually*. Each of those little cells has its own beating power. Myocardial cells, or heart cells, will keep beating for as long as they are alive, even if you separate them from the heart and put them into a petri dish.

So remember to respect your heart and serve it right, which means giving it loads of nutrition and all of the exercise it needs to stay pumped and ready. Your heart may not be heart-shaped, and it may not be cute. But it is BEAUTIFUL and it is your job to take care of it.

BRAIN FOOD

Oxygen is brain food. And since exercise can increase the amount of oxygen delivered to the brain by the blood, you should want to exercise more. In fact, recent research has shown that when kids exercise, it benefits their cognitive performance and may boost learning capacity.

Your brain, which cannot survive without oxygen for more than a few minutes, is the centre of your nervous system, housing your memory, your intelligence, your ability to reason. It has up to a trillion neurons, which are interconnected nerve cells that can transmit information that you interpret as pain or desire or joy or a million other things. And no matter how many languages you've learnt or recite the succession of kings and queens, you use only a small percentage of your brain's total capacity.

The brain is an astounding and mysterious organ. It has been dissected, cross-sectioned, bisected, put in jars, put on slides and put under microscopes. It has been analysed and poked and sliced and diced and discussed and then poked again. And yet nobody has ever seen a personality or a hope or a thought. What we have seen is the way neurons communicate with one another to create feeling, store information, and control behaviour, and the way they send signals that translate into thoughts, emotions and impulses.

Your brain is the centre of your nervous system. Your nervous system controls everything in your body. Your brain connects to your spinal cord, and together they form your central nervous system, which in turn connects to miles and miles of nerves, which conduct (wait for it) *ELECTRICITY* (cool, right?) and send signals from your brain to your arms, your legs, your gut, your toes, your tongue and the little part on the back of your neck that gets tingles when you feel afraid. All of those nerves comprise your peripheral nervous system.

Your peripheral nervous system is made of two systems, one that you consciously control and one that acts without you even knowing it.

Your *involuntary nervous system* oversees processes that occur without you having to think much about them, such as sweating, digesting, peeing and sexual arousal.

Your *voluntary nervous system* oversees all of the movement you do intentionally, which is movement controlled by your motor neurons. This includes everything from using a fork to typing on your computer (as I'm doing right now) to training your body.

So when you run a mile and start to breathe heavily, your nervous system is operating along two channels: the voluntary one that laced up your trainers and got your body on the trail and the involuntary one that upped your oxygen delivery just when your muscles needed it most.

Your muscles need more oxygen when you are moving than they do when you are resting. Nutrition is fuel for your workout, and so is oxygen. Carbon dioxide is waste, and so is sweat.

Your heart and lungs work together to get your muscles the oxygen they need to turn nutrients into ATP. As you train and you begin to breathe at a more rapid pace, your lungs are absorbing more air than usual. Your lungs are expansive and designed to process much more oxygen than humans actually need when at rest. This extra potential is useful when you begin training, because as you exercise and your body requires *more* oxygen to keep you going, your lungs can work harder, getting rid of the excess carbon dioxide produced by rapid and/or sustained movement and giving you more oxygen as you huff and puff your way up that hill.

Meanwhile, your heart starts to beat faster. As a lady, when you are just chillaxing, your heart rate is usually between seventy-two and eighty beats a minute. When you're training or your body is under some kind of stress, your heart rate increases. When your heart rate increases, more blood can be pumped around the body, which can provide your muscles with oxygen and the other nutrients you need to fuel your physical activity (a response that came in handy in those 'run away from the lion around the corner' days).

Because you are increasing the workload of your energy pathways, you are generating more waste. So more blood is pumped during exercise to deliver nutrients and oxygen to your muscles and eliminate waste products from the body. What you will see is an increase in breathing when you start to exercise that may even continue after you have stopped exercising.

The more you train, you will find that your breath returns 'back to normal' more quickly – that is one way to tell that your training has improved your circulatory and respiratory systems. Meanwhile, your ability to go farther even faster is evidence that your muscles and bones are also benefiting from your healthy human exertion.

CHAPTER 18

STRUCTURAL SUPPORT

H AVE YOU EVER SEEN a jellyfish? It looks like a bowl of jelly because that's basically what it is. Without bones, it has no real form, just a gooshy mass that shape-shifts in the currents. But you have bones, and so instead of a shapeless mass, you have clearly defined arms and legs and a torso and sexy collarbones.

That's just one of the things your bones do for you: they support the soft tissues of your body and give your frame its structure. Your bones also help you move, giving your muscles something to hang on to. And your bones build your blood. Bone marrow, the innermost core of your bones, is where most of your blood cells are formed.

How amazing is that? Your bones aren't just structural, aren't just about movement – they are also a factory for your blood cells. In addition, your bones are in a constant state of flux. Bones and muscles are living tissues, and they grow and change throughout your life because, just like the rest of you, they are subject to metabolic processes. Bones and muscles are not just inactive, static materials that stay the same forever; they are always being built up or being torn down. The cells in your bones are constantly renewed, and so are the cells of your muscles and your skin. Bones develop in three stages:

BONE GROWTH AND BONE MODELLING. Both of these stages start when we are a tiny foetus and are ongoing until we are teenagers. The size and shape of our bones is determined during these stages, with 90 per cent of your

The human body is always changing. Every day, your body loses billions of cells (we lose about a million skin cells a day alone!) and makes new cells to replace them. Some cells divide and renew, while others quietly die. It's all part of an incredible balancing act that is constantly taking place within your nerves, your muscles, your bones and your organs.

peak bone mass reached by age 18! This means that the size and shape of our bones don't change too much after our teenage years (which is why you won't grow taller in your twenties or thirties).

BONE REMODELLING. Throughout adulthood, a continual process takes place in which existing bone material is broken down and re-formed as new bone to maintain our bone mass.

BONE BY BONE

There are twenty-seven bones in each of your hands, and twenty-six in each foot. Your face has fourteen bones. Your spine has thirty-three. The longest bone in your body is your femur, your thighbone. The smallest is the stapes, in your ear.

All told, as an adult human, you have 206 bones. I say 'adult human', because when you were born, you had closer to 300 bones. Over the years, your bones hardened and some merged together until you developed the skeleton you have now. And all of these bones, from your skull to your spine to your hips, are made of a mixture of minerals (mostly calcium and phosphorus) to provide hardness and protein (collagen) to provide strength and flexibility.

SKULL: The cranium is really made of twenty-two bones, not just one, and the bones are bound together by fibre. When you were born, your skull was made of soft connected plates so that it could fit through the vaginal canal. When you were about one and a half years old, your plates fused together. And because your skull protects your brain, a little hardheadedness here goes a long way.

SPINE: Your spine has four natural curves, like two S's stacked together. The first curve – to the front – is at the neck, where seven of your thirty-three vertebrae make up your cervical spine. The second curve – to the back – is made up of the twelve vertebrae of your thoracic spine, or your upper back. The third curve – to the front – is made of the five vertebrae of your lumbar spine, or your lower back. The final curve – to the back – is the top slope of your butt, the five vertebrae of your sacrum and the four tiny bones that make

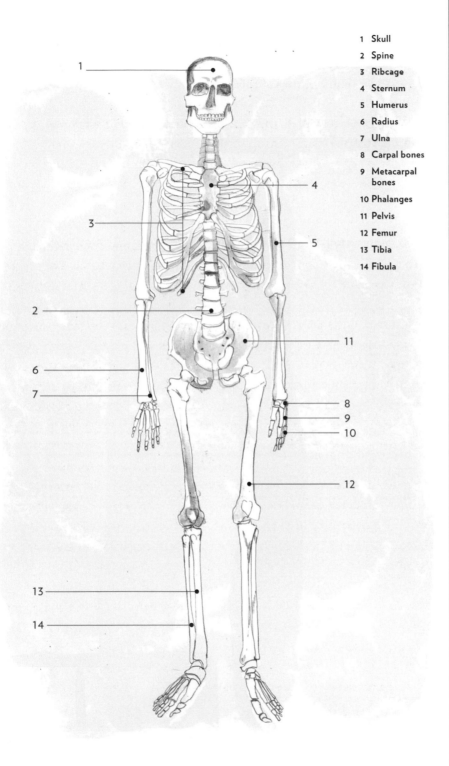

1 Skull
2 Spine
3 Ribcage
4 Sternum
5 Humerus
6 Radius
7 Ulna
8 Carpal bones
9 Metacarpal bones
10 Phalanges
11 Pelvis
12 Femur
13 Tibia
14 Fibula

up the coccyx, your tailbone. Between each bony vertebra and the next is a soft spongy disc that allows you to jump and dance by cushioning the vertebrae and absorbing some of the shock caused by your movements.

CHEST: Your ribcage is a beautiful architectural structure that keeps your heart and lungs safe. You have twelve pairs of ribs connected to the twelve vertebrae of your thoracic spine and held together in front by your sternum, the thick bone that you can feel in the centre of your chest.

Your ribs are also involved in breathing. *Take a big breath in.* Your muscles and diaphragm lift your ribs, which lets your lungs expand and fill with air. *Breathe out.* Your ribs move back down, helping your lungs get out all of the air, getting rid of waste – carbon dioxide – so the process can begin again.

ARMS: Your upper arm has one bone, the *humerus*. Your lower arm has two, the *radius* and the *ulna*. Hold your arm palm up. The bone you feel on the pinky side of the hand, down the back of your arm, is the ulna. Your radius and ulna meet your *carpal bones* in your wrist. In your palm, you have *metacarpal* bones. Your metacarpal bones connect to your *phalanges*. Wiggle your phalanges. (*Phalanges* is just a fancy word for fingers.)

PELVIS: The word *pelvis* comes from the Latin for 'basin'. Muscles, connective tissue and bones rely on the pelvis for guidance and support, and cheerleaders and belly dancers know how to swivel it for maximum effect. The pelvis is the foundation of your lower body; it protects your digestive organs and plays a major role in childbirth.

LEGS: Like your arms, your legs have one large bone at the top and two smaller bones at the bottom. The bone in your thigh is called your *femur*. The two bones in your lower leg are your *tibia* and *fibula*. The larger, the tibia, is your shinbone; the smaller, the fibula, your calf bone, runs down the outer part of your lower leg.

TWO TYPES OF BONE TISSUE

There are two distinct types of bone tissue. **Cortical bone** is very dense and makes up about 80 per cent of your bone; it can be found within your long arm and leg bones, as well as on the outer surface of all your bones. **Trabecular bone**, the remaining 20 per cent of your bone mass, is much more porous and looks like a honeycomb. Trabecular bone is found inside the ends of the long bones, the spine and the pelvis. Most bone fractures commonly occur in trabecular bone.

Your more flexible trabecular bone has a much faster turnover rate than cortical bone, making it more sensitive to changes in nutrition and hormone fluctuations. (That's why calcium is in our list of minerals in Chapter 10: because it's literally what your bones are made of, and eating calcium-rich foods replenishes your calcium.)

BUILDING HEALTHY BONES

To keep your bones healthy, you've got to focus on three things: nutrition, exercise and hormones. That's because sometime in our twenties, our bones achieve their peak bone mass. That means that until your mid- to late-twenties, if you are healthy and not underweight or malnourished, you may be still building more bone than you are breaking down. But if you are underweight or malnourished or if you are missing your periods, there can be a problem with healthy bone formation, and more bone may be being broken down than is being built.

That is why it is so important for women in this age group to build as much bone as they can by eating well and working out. I'm talking to you, Miss 25-Year-Old, rolling your eyes! You have an opportunity to create the highest peak bone mass, the point at which your bones are at their maximum mineral density, right now. You have a chance to lay the best possible skeletal foundation for the rest of your life.

When women are around forty or fifty years old, the changes in our hormone levels that accompany perimenopause and the menopause cause changes in our bones. As a result of this natural shift, women's bones begin to break down more rapidly than the body can rebuild them, resulting in a loss of bone density, which means lighter, weaker, more brittle bones.

The best way to prevent bone problems later in life is to embrace good eating habits, engage in weight-bearing physical activity and stay aware of the regularity of your menstrual cycle, which is a sign that your hormones are functioning normally.

When bone mass declines to a point where the bone becomes very porous and fragile, you may be diagnosed with osteoporosis, which is a weakening of the bones. When you have osteoporosis, it is easy to break a bone and become seriously injured by even a minor slip or fall. That's why it's so important to build a foundation of healthy, strong bones while you still can!

I feel fortunate that I began training when I was twenty-six. At the time, I knew that building bone mass was one of the benefits of working out, but now it means more to me than ever because I've been building bones for the last fifteen years consistently.

Working out builds muscle, and it builds bone. The key is finding those exercises that load the bone with weight, which are called weight-bearing exercises. Many forms of activity that stress the bone and muscle in a helpful way will help maintain our bone mass. These weight-bearing exercises occur when you are upright and moving against gravity. Exercises like jumping rope, running, or just plain dancing and jumping around all put stress on your bones and are considered weight-bearing exercises. Activities like biking and swimming, however, are not considered weight-bearing exercises. When you strengthen your muscles by lifting weights, your muscle contractions put stress onto your bone, causing you to stimulate bone in a different way. Such muscle-strengthening exercises are also important for your bone health and as part of your exercise routine. More stress on your bone gives you stronger bones . . . and the opposite also holds true. So use it or lose it! Loss of bone density for women can start as early as age thirty-five. If you aren't building more bone with consistent movement and training, bone loss will occur. Same goes for your muscles.

Strong bones and strong muscles don't make themselves. They are something that you have to work for. So fuel up and get moving so that you can build those bones and muscles for a lifetime.

CHAPTER 19

MUSCLE
WOMAN

WHY DO I TRAIN? Because muscles are strength and *earning* them teaches us that we can create our own strength. Because connecting with your muscles, and understanding which muscles do what, is a part of being conscious and awake in your body. Because muscles can be flexible, and stretching your body helps you walk more freely and stand taller, and shows you that with effort, you can shift the most basic parts of yourself. Because muscles are power, giving you the tools you need to get there, do it, win it, chase it, own it, whatever it means to you. Because muscles are your personal transport system that get you to your job, where you can earn your keep; to your classes, so you can learn new things; to the airport, so you can go experience new places. Muscles are your heart beating. Your ribcage expanding with every breath you take. They allow you to swim to the edge of the lake. To open a jar without asking for help. To lift and hold a child.

Higher levels of muscle strength are also associated with lower risk of heart disease and chronic diseases, and lower risk of death by any cause. And, people, it is *worth the effort*! Especially because strength training just two or three times a week benefits your overall health *and* gives you a chance to get to know your muscles more intimately.

I promise you this: even if right now your muscles feel like wet noodles, you can do this. You can become strong. You can have arms that feel like power tools instead of pasta. You just have to start moving, which is exactly what your muscles were designed to do .

THE SIX HUNDRED

You have six hundred muscles, and they are the reason you can digest your food, pick things up and put them down, shake your head yes or no and measure out 250 grams of rice. Some of them are under your control; they do your bidding, like when you measure out 250 grams of rice. Some of them do their own thing, like when your body turns that rice into energy after you have cooked and eaten it.

There are four types of muscles:

SKELETAL MUSCLES: Imagine walking across the room, and thank your skeletal muscles. Your skeletal muscles are voluntary – they do what you want them to do. Your skeletal muscles are strengthened by your weight-bearing exercise, strengthening exercise and cardiovascular exercise.

VISCERAL MUSCLES: See how you are breathing quietly while reading? You aren't thinking, *breathe, breathe, breathe,* are you? Thank your visceral muscles, which do their thing independent of your conscious brain. Your visceral muscles are thin sheets that cover your inner organs, and they have secret, silent responsibilities that keep you alive, like contracting your digestive system in rhythmic movements called peristalsis, which is what moves your food from one digestive organ to the next.

CARDIAC MUSCLES: Put your hand over your heart. Your cardiac muscles make your heart beat, and they are found only in your heart.

HYBRIDS: Take a deep breath. The hybrids, like your diaphragm, control breathing unconsciously, but you can breathe more slowly and deeply if you try.

HOW SKELETAL MUSCLES MOVE

Before you get into your training, I want you to understand a few more things about your skeletal muscles and how they support your life. Think about the dolls you played with as a child. I bet they had a very limited range of motion – perhaps their arms moved back and forth but not around and around. Now think about your body. It can reach up and down, back and forth, around and around. How amazing is that? It's especially amazing when you consider all the teamwork that goes into making your structure operate optimally. Because at their basis, muscles are pretty simple. All they can do is contract and then relax.

CONTRACT, RELAX

Make a fist. Tighter! Now let go. When you squeeze your hand into a tight little fist, your muscles are contracting. When you let go, they relax. Contract and relax. It doesn't matter if you're lifting an egg or a boulder; it's the same motion.

Those contractions and relaxations are controlled by two proteins: actin and myosin, which live in the cells of your muscles. When you lift a heavy bag of shopping, the proteins move past one another to help the muscle contract. Once you set the bag down on your kitchen floor, your muscles call the mitochondria into play, awaiting a dose of ATP to help the muscle relax.

Because all muscles can really do is contract and then relax, they are sometimes paired up with muscles that pull and release in the opposite direction, thereby giving your body a much wider range of motion because the muscles can work together to tug your limbs and joints in more than one direction.

FORWARD AND BACK

Here's an example: kick your own ass. Seriously – *kick yourself on the butt with your heel by bending your knee and raising your foot up behind you. Now lower your leg and straighten it completely. Once more.* OK, cool. Here's what's happening when you do that: the muscle that runs down the front of your thigh is called your quadriceps. The muscle that runs down the back of your thigh is your hamstring. These muscles are working as a team to operate your leg. Your hamstring is bending your knee by contracting, while your quadriceps relaxes and lets the hamstring take over. Then your quad straightens your leg by contracting, while the hamstring relaxes and lets the quad take over.

Got it? Bend – compliments of the hamstrings on the back of the thigh. Straighten – cheers to the quads on the front of the thigh. The quads work in opposition to the hamstrings, working together so that you can put your leg down again after you raise it. When one contracts, the other relaxes. By pairing muscles that contract and relax in opposite directions, your body can move forwards as well as backwards, bend as well as straighten.

SIDE TO SIDE

Of course, if your arms and legs moved only forwards and backwards, you'd be more robot than human. So your body has other muscles that provide for different actions so that you can glide, dance, and tumble and not just move like R2-D2.

Imagine taking a pen and drawing a line from the centre of your forehead straight down through your belly button to the ground between your feet: that's called your midline. Your body has muscles that are cleverly designed to move your limbs in towards and out away from your midline.

Move your leg out to the side, slowly: you just used your glutes – minimus and medius – which run down the side of your hip. (The gluteus maximus is the part of your glutes that forms the major part of what you think of as your booty.) Now *swing your leg back towards your midline*: you just used your adductors and your groin muscles to accomplish that action.

IN AND OUT, AROUND AND AROUND

Some muscles are there to move your limbs – your legs and arms – away from your body and then back in, so you can swim the breaststroke or clap your hands. These are collectively called adductors and abductors. Abductors move your limbs *away* from your midline. Examples include your glutes (for your legs) and your deltoids (for your arms). Adductors move your limbs back in. The adductors and abductors work together to move your limbs in and out.

Moving back and forth and side to side is still pretty robotic. So your body has other muscles, the rotators, that help you move your arms and legs in a circular motion so you can wind up for a pitch, turn a skipping rope, or do a really awesome dance move. For instance, there are four muscles in your shoulder that form part of a structure called the rotator cuff; they are the reason your arms can swing so freely. Your hips also have a lateral rotator group that lets you do a fly kick if you feel like it.

GET TO KNOW YOUR MUSCLES

After I did *Charlie's Angels* – after I was introduced to my body – I got to know my body and became committed to my body. Then I got to meet all of my skeletal muscles, one by one. I got to know how they felt when they were tired and how they felt when they were at their prime. How they felt during a training session, and how they felt the day after. And I got to know how I felt when I was taking care of my muscles, when I was strengthening them and stretching them, giving them challenging tasks and feeding them the amino acids they needed to repair and giving them time to rest. And sure, after a workout I might feel sore and achy, but I came to understand that those aches and pains were literally growing pains. They were the kind of aches that were HELPING me. They were the kind of pains that were HEALING me.

Over time, the soreness gave way to a resilience that I began to crave. After my rest days, my body would practically beg to me to keep training so that I could keep having those feelings. So let's get to know all of our glorious muscles on a first-name basis.

ARMS AND SHOULDERS

BICEPS: Flex your arm and squeeze your fist, and there it is: your biceps! Every time you pick up a shopping bag or a bowling ball, you are using your biceps.

DELTOIDS: Put your right hand on your left shoulder. That's your deltoid, a large muscle that helps you lift your arm and move it away from your body so you can do things like hail a cab.

ROTATOR CUFF: The deltoid works with the rotator cuff, a group of four muscles that surround the shoulder joint and work together to stabilise the socket so it can rotate within the joint, thankfully keeping your arm where it belongs. If you're into tennis, you might already know that this is a commonly injured muscle for tennis players, who are constantly rotating their arms as they take those backhand shots.

TRAPEZIUS: Shrug your shoulders. Hey, trapezius! The trap is a big muscle, covering your neck, upper back and shoulders. Because it's so large and close to the surface, it sometimes steps in when other muscles should be working, leading to weakness in areas like the neck, and soreness in the traps. In our modern life, if we're hunched over computers and steering wheels all day, it's easy for the trapezius to get really sore and tight. Doing head rolls can help.

BACK

LATISSIMUS DORSI: The latissimi dorsi (aka your lats) run down your back, one on the left and one on the right, and they are among the largest muscles in the body. Sit down in an armchair and then use your arms to push yourself to standing. Your lats helped you do that. The lats are used to draw the arm inwards and downwards – you use them anytime you do a chin-up or go climbing at a gym. Your lats are powerful (and sexy).

CHEST

PECTORALIS MAJOR: These are your pecs, baby. The pectoralis major is the largest and most superficial of the two chest muscles. When we think about pecs, we usually associate them with guys, but us ladies have them too. They help to hold your shoulders straight across and open to create a beautiful space for a lovely swooping neckline. And they form the first line of defence for your heart and lungs by covering the part of your true ribs that hold your heart and part of your lungs. When you do a push-up, you are working your pecs.

ABDOMEN

RECTUS ABDOMINIS: Lie on the floor. Now do a crunch. Don't worry, just one. You've just activated your rectus abdominus, otherwise known as the six-pack, or the abs. The six-pack muscle is right at the top of your ab muscles, which is why doing lots of crunches results in that distinctive outline. The real job of this muscle is not to look good on the beach but to hold your organs in and to flex your lumbar spine, which is the lowest third of your spine. Having strong abdominal muscles is important because it helps with posture, as well as balance. When you see someone walking a tightrope, when you see Cirque du Soleil acrobats tumbling and supporting one another on their pinky fingers, most of that strength and all of the balance come from their core. Our abdominal muscles work as a corset holding in the centre of our gravity so that we have more control over where our bodies move. It's basically ground control for your movements, providing balance and stability for everything you do – and it's one of the muscles responsible for your beautiful posture.

OBLIQUES: The external obliques are the muscles that run down the sides of your abdomen, and they jump into action every time you bend or twist your body. The obliques have the job of holding your insides inside, where they belong. To strengthen your obliques, try some twisting crunches: lie on your back with your feet on the floor, knees bent. Put your hands behind your head, elbows bent. Now sit up one-quarter of the way, aiming your left elbow at your right knee, then aiming your right elbow at your left knee.

LEGS

GLUTEUS MAXIMUS: The gluteus maximus, medius and minimus make up your glutes, otherwise known as your perky little butt muscles. The gluteus maximus is the biggest butt muscle, right at the pocket of your jeans. The gluteus medius and gluteus minimus run down your hips at the side. The glutes help you walk, run, jump and sit. Tightness in the glutes can be related to a sedentary lifestyle (i.e., too much sitting), which may contribute to back pain. You can stretch your glutes by lying down and hugging

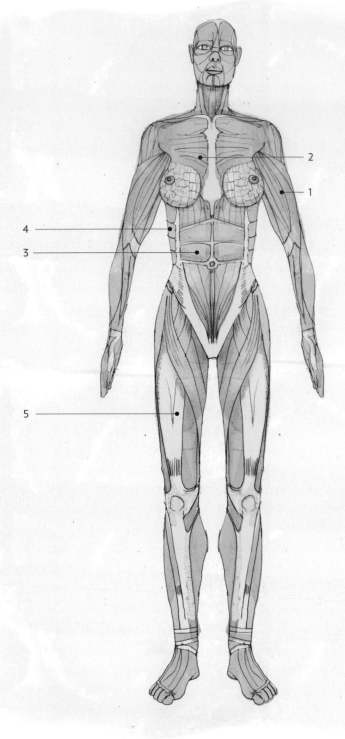

1 Biceps
2 Pectoralis Major
3 Rectus Abdominis
4 Obliques
5 Quadriceps

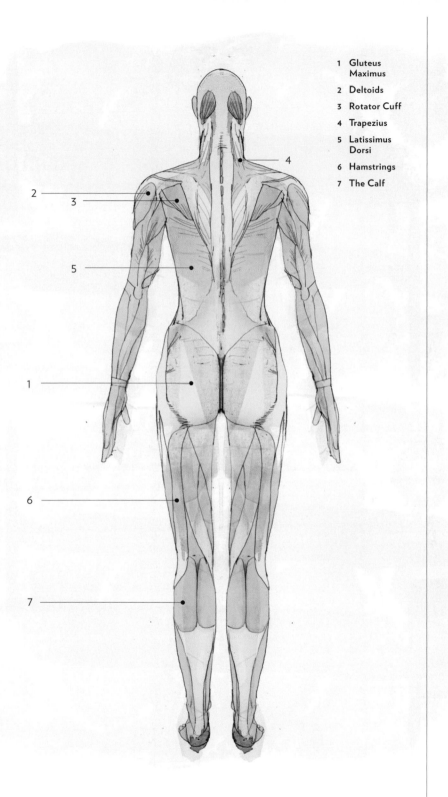

1 Gluteus
   Maximus
2 Deltoids
3 Rotator Cuff
4 Trapezius
5 Latissimus
   Dorsi
6 Hamstrings
7 The Calf

one knee to your chest while keeping the other leg stretched long on the floor, then switching sides.

HAMSTRINGS: Modern living is rough on hamstrings, which can get tight when we sit too often for too long. The hamstrings are located at the back of the thighs. This set of three muscles works the hip joint and the knee joint together to bend your knee and move your leg backwards from the hip. To stretch your hamstrings, try bending forwards at the waist, tightening the muscles on the front of the thighs to release the hamstrings along the back. If your hands reach the ground, great. If they don't, keep your hands on your knees, and don't overdo it. It's best when stretching to let your muscles loosen slowly.

QUADRICEPS: The quads are a group of four muscles found in the front of the thigh. Their main function is to straighten the knee joint from a bent position. The rectus femoris muscle, the biggest of your four quads, also works to flex the hip so that you can lift your thigh up, as when you walk up the stairs. Still don't know where your quads are? Put on a pair of heels. See those sexy muscles that bulge nicely at the front of your thigh? Those are your quads. To engage your quads, try a wall squat. Stand with your back against the wall, then slowly bend your knees as if you are sitting in a chair. Hold for thirty to sixty seconds and release.

THE CALF: You have two major calf muscles, the gastrocnemius and soleus. The bigger and closer to the skin, your gastrocnemius, propels you as you run and walk. When you are standing upright, that's your soleus.

Want to engage your calf? Try calf raises: stand on your tiptoes, then lower; tiptoes, then lower. Those are your calf muscles lifting you up.

BUILDING MUSCLE

Just like with your bone density, the more you build your muscles when you're younger, the more able you are to lay the groundwork for a healthy, strong future.

Building muscle is a choice that you make every day, via the foods you eat and the activities you do. Here's what many people don't realize: anytime a muscle is engaged, you're building strength. When you lift a weight, what builds strength is not just having the weight in your hand. It's that you're squeezing the muscle. The longer you hold that squeeze, the more strength you put into the muscle.

If you were to significantly injure your muscle by sustaining a muscle tear, it would be repaired with groups of the amino acids that you've ingested by eating protein. That repair is like laying bricks and mortar to build a structure; it reinforces the muscle, nature's way of making sure deterioration doesn't occur. It's another example of how the body moves towards strength when you give it plenty of protein and movement.

Squeeze your butt muscles. Yes, right now, while you're sitting. Pull your abs in, too, while you're at it. Hold . . . now release. That's all it takes. Squeeze and contract any muscle in your body, and you can get stronger, even while you're sitting here reading this book. With the recognition that we can build muscle by engaging it anytime, anywhere, all of our movements can become strengthening movements.

When I first transformed my body into a strong body, those feelings of strength started in my muscles, and they crept into my head and my heart, into the way I saw the world and the way I saw myself. And as I got stronger, I began to feel more powerful, like I could pick up anything that got in my way and just set it gently aside so that I could keep moving. As my physical abilities increased, my understanding of what I could do – in my training and in my career and in my entire life, for that matter – grew by leaps and bounds. Grew humongously. Grew exponentially.

That same power can be yours, if you want it to be.

CHAPTER 20

THE BASICS OF TRAINING

———

BODYBUILDING. THAT'S RIGHT. BODYBUILDING. Don't freak out! I'm not talking about building the kind of bulging biceps that will land you on the cover of a muscle magazine. I'm talking about building the actual structure of your body. The architecture.

Your body is a structure that is held together by your bones and muscles. Just like any structure, how well it is built determines how long it will remain strong and upright.

Think about the architecture of a house. If the walls were built at a tilt, wouldn't you be nervous that eventually the house would lean in on itself until it just buckled and caved in? The body isn't that different. If all of the support beams and ties aren't strong and able to bear their load, things start to break down: disks become herniated, knees go out, hips strain, ankles twinge, necks ache, stress fractures occur . . .

Your bones rely on your muscles to help keep them strong, and your muscles rely on you to make them strong. Building your body just means creating strength in all parts of your body in order to create a structure that will be able to carry you through life so that you can be strong and capable. And creating that strength means TRAINING.

Training means real weights, real effort, real sweat. You can't train by just thinking about it. You can't train by just reading about it. But you can think about it, read about it, then do research and find a programme that's right for you.

I can tell you this: if you've never trained, there is a programme out there that is right for you. If you used to train and you stopped for this reason or that reason, there is another programme that will be right for you.

If you're already a fitness expert or somebody who spends a lot of time using and building their muscles, that's awesome! I bet you found your programme, and spent a lot of time and energy focussed on it. And that's the message I need all of you girls who haven't spent hours under the tutelage of a coach or at a training camp or at the gym to understand: it takes time to sculpt a body into the body of a trained athlete. It takes hours and hours and days and days and lots and lots of passion and energy and patience.

So if you have never trained, please be gentle with yourself. Be forgiving. It took your whole life to create your body as it currently is at this moment. It's going to take time to make changes. You are not going to become a fitness expert overnight or in three months or even a year. You can get sweaty right now, this very second. You can feel better in minutes if you just start moving. But you can't become transformed overnight.

It is crucial to take the time to learn how to do things properly. Setting yourself up for success is setting yourself up the right way, and this goes for anything and everything you do in life. There's a reason you start school in reception and not in the sixth form. Before you learn to read Shakespeare, you have to learn the alphabet.

The purpose of this book is to help you connect with your body and understand its basic needs and functions so that you have the knowledge and confidence to go out into the world and be the healthiest person you can be.

The best place to start is to pick an activity that has just the right amount of challenge to push you but will make you want to keep coming back for more. A programme that is too easy will be boring. A programme that is too challenging will be frustrating. Find something that makes you want to keep doing it, whether it's mountain biking, riding your bike to work or riding on a stationary bike.

Once you find your physical outlet, you must also find consistency. And you must stay consistent for the rest of your life. Yep, that's right: some sort of physical activity must be a part of your life for the rest of your life.

Because if you want to keep that body you'd better BUILD that body.

THE RIGHT TRAINING PROGRAMME FOR YOU IS ...

- At your level
- In your neighbourhood
- One that excites you
- One that will make you sweat
- One you will stick to

FORM IS EVERYTHING

No discussion of exercise is complete without talking about form. When you watch gymnastics and see those elastic bodies hurtling through the air, the judges aren't scoring based on how daring or pretty a routine is; they are looking for form, for how well the shape of the movement matches the intention behind it.

There is a way to do any exercise efficiently, in a way that builds your structure, keeps you safe, conserves energy and maximizes movement. Form guarantees that efficiency. Form is the mechanics of a body using its muscles, its abilities, in the right way. It's how a body gets the most out of every movement.

Form is as important to a beginner exerciser picking up a weight for the first time as it is for a professional bodybuilder picking up a weight for the ten-thousandth time. Form is as important in your posture as it is in your walking, running, sitting and training.

BASIC TRAINING

I like to change up my workouts, finding my physical activity in a variety of different places and activities so that I engage my body fully. It keeps me agile and responsive, as well as strong, fit and happy.

Although I like variety, I aim to do some abdominal work in every session because your core is the centre of your strength. It supports your spine, helping to carry the burden of your bodyweight. And when you lift weights, engaging your core strength will help keep you steady and balanced so that you can maintain proper form.

When thinking about working out, there are a couple of different approaches – you can concentrate on exercising muscle groups or isolating specific muscles. You might strengthen multiple muscle groups by doing exercises that engage several muscle groups and joints at once. This would include movements that target the major muscle groups of the chest, trunk, back, shoulders, arms, hips and legs. Or you can do exercises that isolate muscles, such as the abdominals, lower back muscles, hamstrings, quadriceps, biceps and calves.

Either way, it's always a good idea to train the opposing muscle groups (the antagonists) in order to prevent muscle imbalances. Abs and back, hamstrings and quads – those are examples of opposing muscle groups that should be trained simultaneously.

A+ FOR FORM

When you take a group fitness class such as yoga or spinning, make sure you let the instructor know that form is important to you. Tell him or her that you welcome any corrections, and you'll likely get a little more one-on-one attention. If you work with a trainer, make sure that you ask her to concentrate on your form and correct you as often as necessary until you get it right.

Poor form not only means the body doesn't perform efficiently but it's also a sure way to injure yourself. Maybe not the first time, but continuous bad form will eventually take a toll. So before you do *any* exercise, make sure that you've learnt the proper form from a reliable source.

POWER YOUR TRAINING

Before and after your workout, it's important to give your body what it needs to perform and repair: FUEL and WATER. Fuel and water get your body ready for a workout and allow your body to repair and restore afterward.

I drink water before, during and after a workout. And I fuel up before, during and after. By which I mean that I eat complex carbs. If you want to keep your body happy and exercise to your potential, you will need to eat a meal before and after your workout that has plenty of carbs for fuel and protein for muscle building and repair.

When you eat carbs, the energy (glucose) that isn't immediately used gets stored in the liver and muscles as glycogen. Glycogen is your body's energy reserve. When you've burnt through all of the available carbs from your last meal, your body can easily get at that glycogen, powering through those last few miles.

Fuel gets you going and keeps you moving. Water keeps your cells hydrated, cools you down and replenishes the fluids you lose when you get SWEATY. Here's a more specific breakdown of what your body needs to succeed before, during and after exercise.

TWO HOURS BEFORE EXERCISE

WATER: Drink at least 500 millilitres of fluid, preferably water, about two hours before you exercise. Then about fifteen minutes before exercise, drink 125 to 250 millilitres more to top up your body's water stores.

FUEL: Two hours before moderate exercise, like a five-kilometre jog, a spinning class or a hike, eat a combination of carbs, protein and fat. Here's why: when you eat carbs alone, they are digested quickly. Adding fat slows the process, as does protein. And protein helps with muscle repair. So if you just have a piece of whole-wheat toast, it isn't enough. If you just have some avocado for fat, it isn't enough. If you just have some cheese for protein, it isn't enough. But put them all together for a piece of whole-wheat toast with avocado and cheese, and you will have energy for the long haul.

It's also important to know that if you can't fuel up two hours before your workout because you're working out at five thirty in the morning, even a few mouthfuls of porridge or a piece of toast with some almond butter will do good things for you. It can be as little as a half hour before your workout. You don't need a lot, just something to put in your stomach, a little gesture to your body that you are acknowledging its need for fuel. It kick-starts your engine and fires up your metabolism, ensuring that your body will be using glucose, glycogen and fat to fuel your workout, and not tapping into your precious muscles. Never work out on an empty stomach.

DURING EXERCISE

WATER: Drink approximately 175 to 350 millilitres every fifteen to twenty minutes. An average 'gulp' is about 30 millilitres. No matter how hard I work out, as long I drink water, I feel hydrated and replenished.

FUEL: If you are doing endurance exercise, like training for a marathon, you require quick access to ready fuel. Talk to your coach or trainer about what snacks will give you ready access to additional glucose. The amount of fuel required and the type will depend on your total daily energy expenditure, the type of sport and environmental conditions.

AFTER EXERCISE

WATER: Replace fluids as quickly as possible. Having a big bottle of water ready for the end of your session is a must.

FUEL: Many of us feel aches and pains after an intense workout. One reason for that soreness is because your muscle tissue may be repairing itself during a workout so that you can generate new, stronger muscle. During the recovery period after exercise, your body repairs and rebuilds muscle (that's literally body-building!). Having a 'recovery plan' that includes smart nutrition choices can help you recover faster and better.

The recovery window after a workout is about forty-five minutes to an hour, so try to eat a recovery meal within that time frame. Carbs will replenish glycogen stores, and protein will help to repair and rebuild muscle tissue, so an ideal recovery meal includes both carbs and protein (in a four-to-one ratio). Your body is primed to refuel and repair at a higher rate during the recovery phase than if you delay your next meal, so take advantage of this prime opportunity to recover and give your body what it's asking for!

GET YOUR STRETCH ON

Ahhh, stretching. For me, stretching is a morning, noon and night activity. It's a way to familiarize myself with my body, to relax my mind and to get my muscles ready to bear weight when I train.

I find myself stretching throughout the day, even if it's just dropping over to touch my toes while waiting for the elevator, or doing head rolls to stretch out my neck while I'm cooking my eggs in the morning. Stretching goes hand in hand with breathing, so when you take a deep breath and let out a big exhale, you're feeding your body oxygen and releasing toxins into your bloodstream (including lactic acid from your muscles) so they can be whisked away. Breathing and stretching acts as a mini-detox; it helps to clear the mind at the same time as it helps to clear the body. We'll get into more detail about the mind-body benefits of stretching in Chapter 24.

Stretching is also a great way to wind down at the end of the day, because it's essentially a release of built-up energy. I might lie on my floor and just twist and turn my body, breathing in and out. Or I'll bring my knees to my chest and take deep breaths. Or do a seated stretch, and reach forwards to touch my toes. It's like taking a big, full-body yawn. Release the tired energy, bring in the new energy.

Stretching and flexibility exercises that target your major muscles and tendons in your neck, shoulders, trunk, hips, legs and ankles can also contribute to better posture and balance. And whatever your age, no matter how flexible you are, regular stretching can improve your range of motion and flexibility. And that's something you want, trust me. Flexibility tends to decrease as you get older. The good news is that flexibility can be improved. Stretch regularly at least two to three times a week, and you can see improvement in as soon as three to four weeks.

Current research tells us that these flexibility exercises are most effective when performed while the muscles are a bit warm, such as after a workout. Personally, I always make stretching a part of my workout. I usually do a little warm-up on the treadmill and then stretch all my muscle groups before I do any lifting or resistance training.

Before I begin any exercise, I connect with the muscle group I'm about to engage to make sure that there isn't any stiffness. Stiffness could lead to straining or locking up my muscles, which I don't want to do. So if I feel any stiffness, I'll take the time to stretch out that muscle before engaging it. Once I have the OK from that muscle that it's all ready and warmed up, I'll go ahead and start my exercise.

STRETCHING TIPS

Hold the stretch for ten to thirty seconds at the point of tightness. Relax into the stretch; don't push into it. Breathe and release.

Remember that we all have our own level of flexibility. Stretching is not about imitating the girl in your yoga class who can fold in sixths like an origami crane. Stretching is about feeling your way to the places where your muscles have sensation without causing injury.

TAKING CARE OF YOUR BODY WHILE TRAINING

When I was doing *Charlie's Angels,* and Master Cheung-yan Yuen and my trainer Tiger Chen told us that pain would become our best friend, I had no idea just how right-on those words would prove to be. One of the most important things they taught me was the difference between PAIN and INJURY. So let's be really clear: *you do not want to get injured.* You want to get STRONGER.

To do that, you will have to give your body enough water and fuel. You will have to learn how to really focus, how to really connect to your body, in order to understand the difference between feeling the pain of growth, the pain of creating strength – and the pain of actually hurting and injuring your body.

When your body is screaming during a workout, you must ask yourself, *Am I in pain or am I injured?*

How do you know the difference?

INJURY: With injury, there is usually a sharp pain that comes very quickly, almost like a warning alarm from your body. It sends it out just before you are about to really do yourself in, so you need to react as quickly as possible. That's why it is important to be present, connected and paying attention to your body and how you are using it.

PAIN: If you are just feeling the discomfort of pushing the threshold of your strength, where you're struggling to hold a yoga pose for another few seconds or do another bicep curl, try to keep pushing if you can (if absolutely necessary, take a break to recover). This is not the kind of pain you should fear: this is the pain of growth. It's worth challenging your limits.

One of my favourite sayings is 'Pain is weakness leaving your body'. Isn't that such a great thing to know? When you push through to the other side of pain, you are stronger for it; you let go of weakness and build resilience in your mind and body. Because the more you resist giving up when you can actually keep going, the more your mind will also adapt – it will learn that it is capable and it will power you through the pain. It doesn't have to be scared of doing one more rep, because you are connected to your body, and it can trust you to either stop before getting injured or keep going, building strength and capability. It all works together.

Part of training is learning the difference between those two feelings. Once you get it, you can go even further. Once you identify the line, you learn how to cross it safely. Because some kinds of pain are required for growth.

THE SWEATIER THE BETTER

Women don't just glow. We don't just glisten. We SWEAT. As we should, because using our bodies so intensely that we sweat results in amazing things.

One very fun Saturday, I went with my nieces and nephew to a huge warehouselike playland full of trampolines, padded from wall to wall. An entire room full of trampolines! It's like heaven. You just jump from one trampoline to the next to the next to the next. It is ridiculously fun – and everybody gets really *sweaty*. So there we were, just jumping from one trampoline to the next, a hundred fourteen-year-olds and one forty-year-old – me.

If you want to build strength, you must be comfortable with discomfort.

An hour later, we are all laughing, giggling, exhausted and just dripping with sweat. Soaking wet. Drenched. And I looked at the kids and said, "Doesn't it feel good to sweat?"

They all said, "YES!!"

I said, "I do this every day."

They were flabbergasted. "*Every* day?"

And I said, "Yes. Every day."

"What do you do?"

"I run. And I hike. And I strength-train."

When I dropped them off back at home, they showed me a hill near their house. And they said, "If we run up and down that hill, will we sweat?"

And I said, "Yes! Yes you will."

And they were so excited. Because they had connected with how amazing it feels when we give ourselves opportunities to sweat. To move. To liberate our bodies from chairs and cars and cubicles and the couch. To just let our arms and legs fly. To let our bodies do what they were designed to do.

I love to sweat. I love to train. I love to build new muscles and feel the power they give my body. Unless I'm taking a day off for rest, I really do sweat every single day. Sweating is FUN. Moving and sweating are what we women were made to do.

When you get going with the free weights and your face starts to feel warm and that circle of sweat appears on your sports bra, when you lace up your kicks and hit a dirt trail and suddenly your entire body is covered with a thin film of sweat, that's when you know it's working. Sweating is how you know you're doing something RIGHT.

CHAPTER 21

YOUR
LADY BODY

—

HEY, GIRL! WE'VE SPENT a lot of time over the past few chapters talking about your human body, and the nutrition and movement you need to keep your human body healthy and strong. In this section, we take a little trip south. Because your body isn't just a generic human body. It's a female human body. A glorious, wonderful, beautiful woman's body. And as such you have needs and parts and hormones and cycles that you'll want to understand if you want to be healthy.

GETTING INTIMATE WITH YOUR INTIMATES

Being a lady is about a lot more than just wearing a bra and bleeding once a month. Your lady body is a beautiful, intricate, fascinating part of your humanness. It is erotic, and it is functional. It gets rid of waste and reproduces and nourishes new little bodies. It can make your toes curl with pleasure, and it can inspire other people's toe-curling responses. Your vagina is an incredible, lovely place, an inspirer of art and desire, and the gateway through which human life can enter the world.

But how much do you *really* know about your lady parts? About when your cycle starts, and what ovulating really is, and how easy or hard it is to get pregnant? About the difference between your uterus and your cervix, your vagina and your labia, your oestrogen and your progesterone and their roles in how your body functions each and every month? Luckily for you, I got some great information from Dr Diana Chavkin, an obstetrician-gynaecologist (ob-gyn) who gave me the lowdown on what's going on down low. Thanks to her, we can connect all of the dots – how nutrition and body weight affect your period, why your vaginal secretions change over the course of a month (oh come on, I know you've noticed) and how staying fit can make sex more fun (whoo-hoo!).

Your lady parts are neatly classified by the medical system as external and internal genitalia, aka the parts that are on the outside and the parts that are on the inside. The external genitalia include the clitoris, labia majora, labia minora and Bartholin's glands (not pictured). The internal genitalia are the vagina, uterus, cervix, Fallopian tubes and ovaries.

And here's the craziest part about how little some of us know about our own bodies – do you know that what you see when you look down, what you probably call your vagina, isn't even really your vagina?! It's your labia majora! So have a look at the illustration below, or grab a mirror and play along.

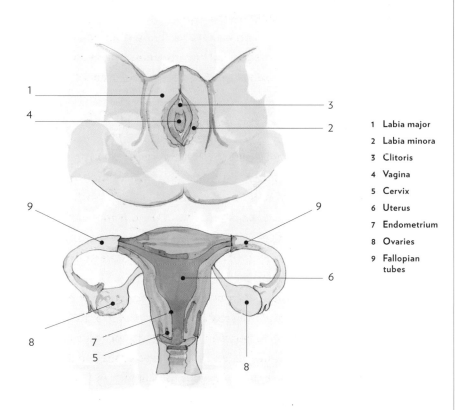

1   Labia major
2   Labia minora
3   Clitoris
4   Vagina
5   Cervix
6   Uterus
7   Endometrium
8   Ovaries
9   Fallopian
    tubes

LABIA MAJORA: When you glance down at the area you collectively call your 'vagina', the parts that are covered with hair (if you haven't waxed it all off) are the labia majora.

LABIA MINORA: The labia minora live just within the majora, and, like nipples, every woman's are unique. They come in different shapes, sizes and colours, all equally as beautiful as the next. Within the labia minora, you will find the openings to the vagina and the urethra.

BARTHOLIN'S GLANDS: These glands are located beside the vaginal opening and secrete a mucus that helps to lubricate your vagina.

SKENE'S GLANDS (NOT PICTURED): These glands release fluid during sexual stimulation, and they are related to sexual pleasure.

CLITORIS: At the top of the labia minora is the famed 'man in the boat', that most sensitive of sensitive spots on your body that young men throughout the ages have struggled to appease and delight.

VAGINA: The vaginal opening, contained within your labia minora, leads to the vagina, a canal that leads to your cervix.

CERVIX: The opening of the uterus. During birth, the cervix completely 'dilates', or opens up, so a baby can emerge.

UTERUS: Also known as the womb. The uterus is the future apartment for your foetus (should you choose to open it for occupancy).

ENDOMETRIUM: The inner lining of your uterus. Each month that your body is not pregnant, your endometrium thickens and grows. If you do not get pregnant, then this lining is lost and released from your body, which is why you get your period. (More on this later.)

Did you know that all of the eggs that you will ever ovulate, or prepare for fertilization, are already in your ovaries? Every egg that you have or have ever had has been in your body since you were a foetus developing in your mother's body.

OVARIES: Your ovaries are small glands shaped like ovals, one on each side of the uterus. They house all of the eggs that you were born with, and they also produce hormones.

FALLOPIAN TUBES: Your eggs mature in the ovaries and then get sucked up by the Fallopian tubes at the time of ovulation – when an egg emerges for possible fertilization. Your Fallopian tubes are where sperm meets egg, where fertilization takes place. A few days later, the fertilized egg, the embryo, travels to the uterus and makes itself at home.

IN PRAISE OF PUBES

I hear that there's a big fad these days of young women undergoing laser hair removal on all of their lady bits, so I just want to give you a few facts about the lovely curtain of pubic hair that surrounds that glorious, delicate flower of yours.

As far as we know, removing pubic hair offers no medical benefit. All healthy women develop pubic hair as they age, and it likely has some evolutionary advantage since, *we all have it*. There is a lot of speculation as to what those benefits may be . . . like, pubic hair may help protect you from chafing during sex. Or pubic hair may hold our pheromones, the personal scents that make us so sexy to our lovers. We do have medical evidence that removing pubic hair can lead to issues like infections and ingrown hairs. You may have increased risk of some STDs and other skin diseases, because less hair equals more skin contact, potentially exposing you to other people's diseases.

Personally, I think permanent laser hair removal sounds like a crazy idea. Forever? I know you may think you'll be wearing the same style of shoes forever and the same style of jeans forever, but you won't. The idea that vaginas are preferable in a hairless state is a pretty recent phenomenon – and all fads change, people. All fads change.

Pubic hair also serves as a pretty draping that makes it a little mysterious to the one who might be courting your sexiness. Pubes keep the goods private, which can entice a lover to come and take a closer look at what you have to offer. Also, let's be honest: just like every other part of your body, your labia majora is not immune to gravity. Do you really want a hairless vagina for the rest of your life?

It's a personal decision, but I'm just putting it out there. Consider leaving your vagina fully dressed, ladies. Twenty years from now, you will still want to be presenting it to someone special, and it would be nice to let him or her unwrap it like the gift that it is. (Of course, a little grooming never hurt anyone.)

BOOBS

If you don't already see a gynaecologist once a year for an exam that may include a cervical smear and a breast exam, start now. I have a friend who makes an appointment every year around the time of her birthday so she doesn't forget. And really, how perfect is that? She's giving herself the gift of health and possibly helping to save her own life as the years go by.

From time immemorial, our beautiful breasts have been a source of nourishment, prepubescent wonderment, artistic rendering, erotic devotion and a whole lotta magazines on the top shelf. Some boobs are huge pendulums, while some are saucy little apples; some beg for holsters that are practically architectural while others are happy in a barely-there cotton bandeau. Whatever they look like, whether your favourite sports bra has one chamber or two, your boobs have some basic stuff in common with all the other boobs in the room. Whatever the shape, whatever the size, boobs – or mammary glands – are modified sweat glands whose primary biological function is to create milk for our young. Yes, they are sexy. Yes, they look great in a plunge halter. But at their most basic, they are part of the human need for food. Your mum's boobs fed you. Yours will feed any children you have. Sorry, any guys who may be reading: they're not there just so you have somewhere to look when we talk to you.

Imagine your breast as three concentric circles. The largest circle is the breast itself. The next circle is your areola, that light pink to dark and dusky rose holder of your nipple, the littlest circle in our bunch. Within the breast itself are fat, connective tissues and lymph tissues, as well as a layer of muscle underneath.

Although research has not shown that breast self-exams improve survival in women who develop breast cancer, it is important for women to get to know their breasts. Breasts change during the course of the month, so becoming familiar with this cycle will help you to identify any changes that might develop in your breasts.

Women younger than thirty often have 'lumpy bumpy' breasts due to normal hormone changes. These lumps can change throughout the menstrual cycle and go away after your period. If a lump does not resolve after one or more menstrual cycles, then a GP should be consulted.

Women older than thirty who feel a breast lump should see a GP immediately for a breast exam and mammogram. Often, an ultrasound or a biopsy is needed to determine whether the lump is malignant (cancerous) or benign.

YOUR PERIOD

When you first met your 'Aunt Flo', I bet you were really excited and maybe just a little bit freaked out. Since then, you've had to learn how to use the plethora of pastel-packaged products that consume a whole aisle at the chemist's. You might have started using birth control, or not. You may be a virgin. You may have had the experience of being a few days late and spending some quality time worrying about various outcomes. You may have peed on a stick and jumped for joy at the double line that appeared (or sighed with relief when it did not).

The word *menstruation* comes from the root *menstru-*, which is Latin for 'monthly'. You may have heard that a period is 'usually' twenty-eight days long. In fact, most women have cycles that last between twenty-four and thirty-five days, and 20 per cent of women during their reproductive years have a cycle length outside this range. (So it is NORMAL if your periods are not exactly twenty-eight days. But if they're *too* far apart, it can be a sign of a health issue. More on this later.) You don't count your cycle on the calendar according to month and day. You count it according to *your* rhythms. Your menstrual cycle count begins on the first day of full flow (not spotting) of your period. And all of that counting and activity revolves around one very small, very significant determinant of the human race – the egg.

ALL ABOUT YOUR EGGS

Every month, about two weeks from the first day of bleeding, your ovary will release a mature egg that is ready for fertilization.

When you were a twenty-week-old foetus, you had six million eggs in your ovaries. At birth, right when you started crying and your mother began to laugh with joy, you had about one million eggs inside your tiny baby body. By the time your breasts grew, you had about three hundred thousand. Until your late twenties, you have many, many eggs, and are generally very fertile. Fertility is always in decline, because you are constantly losing eggs. This happens through ovulation and because eggs just die off on their own over time in a process called apoptosis, a natural and healthy part of life.

Most women in their early thirties will begin to see a decline in fertility – the ability to become pregnant – which becomes more pronounced in their late thirties and early forties because fewer eggs are present in the ovaries. By the time you stop menstruating, when you are in the menopause, you will essentially be out of eggs.

But before you get nervous about not having six million eggs on standby, remember that only about three hundred to four hundred of *all* of your eggs will ever be called into potentially active duty. You've got a large army in there, but only a select few will go through the cycle with you.

SURFING THE CRIMSON WAVE

Your menstrual cycle has three phases: the follicular phase, which gets eggs ready; the ovulation phase, which gets eggs into the Fallopian tubes; and the luteal phase, when the endometrium is shed. The cycle is regulated by our hormones, including follicle-stimulating hormone (FSH), luteinizing hormone, oestrogen, and progesterone.

Every month, your body prepares a comfy nest by lining your uterus with a receptive endometrium, just in case fertilization takes place. When there is no fertilization and therefore no need for a fluffy uterine home, your body clears the space, shedding the lining meant for the fertilized egg and making space for next month's potential lodger.

That clearing out, that shedding, is when you start your period.

THE FOLLICULAR PHASE

Your eggs are microscopic. They are the largest cells in the body, but they are still just cells, not visible to the human eye. Each egg is covered in a structure called a follicle.

From the development of your eggs to your period, the whole story of your cycle is part of a delicate hormonal dance. Here are the key players:

OESTROGEN: Your period begins with a decrease in your levels of oestrogen and progesterone. That decrease causes the shedding of the endometrium.

You'll know this is starting when you spot blood on your knickers (but hopefully not on your white jeans).*

FSH: With less oestrogen floating around, FSH steps up to the plate. FSH is usually suppressed by oestrogen, so the beginning of your period allows FSH to get in the game and take action: recruiting the follicles to produce eggs. FSH stimulates the development and growth of twelve or so eggs, to get the cycle started for the coming month.

During the follicular phase, the follicle that is the most sensitive to the FSH will grow the most, and it will produce oestrogen. The appearance of oestrogen tells the brain that the FSH has done its work, the follicles are operational, and FSH production can slow down. Now that there is less FSH, only the most sensitive follicle can continue to thrive. The other follicles can't survive, and they die off, making way for one egg to win out (hopefully the most fit). This is why usually only one follicle matures and one egg is released during every ovulation (except in the case of fraternal twins), and it is one of the amazing balancing acts your body performs to keep things working the way they are meant to.

The follicular phase usually lasts around two weeks but can be longer or shorter in many women. This is the phase when length really varies from woman to woman, which is why each of our cycle lengths are different.

THE OVULATION PHASE

Around two weeks after the first day of your period, the follicles and the eggs are ready. When the oestrogen level reaches a threshold, your level of luteinizing hormone increases, causing a chain of events that eventually release the egg from the surface of the ovary into the abdominal cavity, where the Fallopian tube sucks it up like a hoover.

You may have noticed how your cervical mucus changes over the course of your cycle. Right after your period, you have thin mucus. As your hormone levels shift, as your body readies itself for potential pregnancy, your mucus also changes. Just before ovulation, as oestrogen levels increase, your cervi-

---

* If there is a bloodstain, use cold water, not hot, and wash as soon as possible. Hot water makes the proteins in the blood set and that makes the stain stick around longer. Just a little tip from me to you!

cal mucus is designed to help the sperm penetrate and get through the cervix. It can be like egg whites, thick but not too thick. If you had unprotected sex just before ovulation, the mucus would help the sperm find their way up to your mature eggs. Just another one of Mother Nature's ingenious tricks.

THE LUTEAL PHASE

Right after you ovulate, the now empty follicle transforms into something new: the corpus luteum, which secretes progesterone, which gets your uterus ready to set up a home for our hypothetical fertilized egg. After ovulation, the cervix starts to produce a more hostile type of mucus that actually prevents sperm getting through. It is normal for mucus to be very thick at this stage (the effect of high levels of progesterone).

Once released, an egg can be fertilized for about twelve to twenty-four hours, but there are some reports of fertilization occurring up to thirty-six hours later. Sperm can live in the female genital tract for about forty-eight to seventy-two hours. Just because you may not be ovulating when you have sex doesn't mean you can't get pregnant! An egg could be released after intercourse and meet a sperm that's been hanging out, waiting for that egg to drop for two or three days (which is pretty amazing, if you ask me).

Pregnancies have occurred from having sex just once, from six days before to three days after ovulation. (Of course, those are the extremes.) Most pregnancies result from having sex within the three days prior to ovulation.

But if your little egg is protected by a condom, an IUD, the pill, the patch, an arm implant, a Provera injection, or a good old-fashioned "No thanks, I'll sleep at mine tonight", it dies, and the uterus begins to shed its lining, which is when you get home from dinner and realize that your gorgeous new lace knickers have been stained.

HOW BODY WEIGHT AFFECTS YOUR PERIOD

When your periods become irregular – or disappear entirely – it can be a sign that something may be off-kilter with your hormonal environment.

Missing periods is a serious thing. And irregular bleeding does not always mean a normal period. If you miss a period, there are a few possible suspects.

If you're sexually active, pregnancy is a possibility. But after that gasp and a dash to the chemist for a pregnancy test, what if you aren't pregnant? What if your period is just . . . missing?

PCOS

One common reason that women don't get their periods is PCOS (polycystic ovary syndrome), a syndrome that is often related to, and can be made worse by, obesity. We've talked about many of the ways that obesity can harm your health, and here's another: your reproductive health is also affected by excess weight.

Women with PCOS may have many eggs in their ovaries, but they don't get released. No release means that oestrogen just keeps thickening the endometrial lining. That means that a woman with a non-ovulatory cycle can go four to eight months without a period and then bleed heavily because the lining of the uterus became too unstable. The bleeding can last for up to a month and is not considered a regular period because it is not related to an ovulatory event.

Being obese can put women at a greater risk for anovulatory cycles and irregular bleeding. Not only does PCOS negatively impact fertility; it also increases the risk of developing endometrial cancer. If you don't get regular periods and suspect you may have PCOS, see your GP.

NUTRITION

If you are a young woman who has recently lost a lot of weight, or if you are a young woman who works out a lot and you aren't eating enough of the right foods, a missing period is often due to an energy deficit: you spent a lot more than you took in, and your body doesn't have the resources to maintain a regular cycle.

It's important to remember that your period isn't just about reproducing; a regular monthly cycle is a sign that your body is healthy and functioning properly. If you aren't ovulating because you are too thin or are exercising too much, then your ovaries are not producing oestrogen, which has serious health consequences. Not having enough oestrogen can lead to poor bone health and other side-effects. According to Dr Aurelia Nattiv, a team physician at UCLA and a medical representative for U.S.A. Gymnastics and other national athletic governing bodies, missing periods due to excessive training may mean that you have components of the female athlete triad. The female athlete triad is a combination of three things: an energy deficit, whether or not you have an eating disorder; an irregular period; and low bone-mineral density and/or predisposition to stress fractures.

Once bone density is lost, getting it back is difficult. Having enough energy to resume your periods will mean that your body is in more of an energy balance and will produce more oestrogen, which will help preserve bone density. This means that if you are underweight, you may need to gain some weight to get your periods back – and you should WANT to. If you've gone for more than a month without having a period and you've recently lost a lot of weight – see your GP.

THE CRAVINGS CONNECTION

Just before you get your period, you may notice an increased desire for rich chocolate morsels, double helpings of pasta and everything with a scoop of ice cream.

These cravings are different from hunger. Unlike real hunger, which, as we've discussed, is sparked by your cells' need for fuel, cravings don't have much to do with nutrition. Otherwise, we'd have a hankering for kale or brussels sprouts instead of pizza and ice cream. The physiology behind cravings is complicated, and psychologists and other researchers still aren't exactly sure how they work.

But ladies, if you go carb-crazy once a month, you're not alone! An MIT study tracked normal-weight volunteers' snacking patterns during the first days of their cycle and weeks later when they had PMS (premenstrual syndrome). The women ate around an additional 1,100 calories in snacks when they were PMSing! More good news: once PMS passed, any resulting weight gain slipped back off.

What does this mean for you? If you're trying to lose weight and you don't want cravings to derail you, have a snack that satisfies your body's needs in a healthier way. Instead of potato chips, try roasted sweet potatoes drizzled with olive oil and sprinkled with sea salt, or avocados sliced with tomatoes, lemon and coriander, to give yourself maximum carbs and fat to satisfy those cravings. My favourite way to satisfy the salty-crunch craving is with a big bowl of homemade popcorn.

PMS

Magazines talk a lot about PMS, but just because you have your period doesn't mean you *always* get symptoms, and just because you experience symptoms or you're moodier or snackier than usual doesn't mean there is anything wrong with you.

The term *PMS* refers to a group of symptoms that affect you physically and behaviourally in a pattern that occurs during the second half of your menstrual cycle if you are not on birth-control pills. Basically, before you get your period, you feel like crap. Throughout your cycle, your body tissues can sense when levels of oestrogen and progesterone change, and those changes can affect mood-influencers like serotonin.

We don't know why some women get all crabby and anxious and bloated and some don't. The most likely explanation is that those of us who feel it more severely are more sensitive to hormonal shifts.

- Mild PMS affects 75 per cent of women with regular menstrual cycles. Symptoms can include fatigue, bloating, irritability, anxiety, tearfulness and changes in appetite.
- Severe PMS is experienced by only a small percentage of women. Also called premenstrual dysphoric disorder (PMDD), it is characterized by symptoms that include anger, irritability and tension that interferes with daily activities. PMDD affects 3 to 8 per cent of women. If you feel like you might be experiencing PMDD, see your GP.

THE PREGNANT VAGINA

When you're pregnant, you go through a lot of obvious changes, and so do the bacteria in your vagina. We already talked about where your first dose of bacteria came from if you were born vaginally: your mother. Same goes for any children you may have. All mammals develop within the uterus without any bacteria, and as they exit the birth canal on their way to their world debut, they pass through the vagina, which is full of lactic acid bacteria, the kind of bacteria that metabolize milk. The vaginal secretions coat the baby's face and mouth, introducing a number of lactobacilli – the same bacteria that help humans digest breast milk – into the baby's body. How amazing is that? Mammals who need bacteria to digest milk have mothers whose vaginal canals have the bacteria their children will need to get their fuel and survive.

That's one of the reasons that scientists like Dr Maria Gloria Dominguez-Bello are realizing that C-sections and vaginal births are not created equal. The C-section is an amazing operation when it saves the mother's life. But when it is scheduled for convenience, babies miss out on their first dose of incredibly important bacteria, the bacteria that get their microbiome going. Studies have linked C-sections and the lack of healthy bacterial colonies to a rise in allergies, asthma, type 1 diabetes and childhood obesity.

I visited Maria Gloria at her NYU office, where she explained a new procedure she's been working on with healthy mums-to-be who need to have C-sections. Before the woman delivers, a piece of gauze is placed in her vaginal canal, absorbing the bacteria that her baby would have encountered on his or her trip down the canal. After the baby is delivered by C-section, the baby's face and mouth are swabbed with the gauze, thereby delivering all that bacterial goodness, restoring what nature always intended in the first place.

Seriously, for a week after learning that, I kept talking about how amazing the vagina was everywhere I went. At a dinner party, where some guests couldn't believe I was using the word *vagina* at the dinner table. At my sister's house, where my niece had to admit that being a 'vagina face' – i.e., born through a vagina – had its perks, and my nephew wished we would stop talking about it. And at a friend's baby shower, where all the women were into it, because pregnant women have to get really up close and personal with their vaginas.

These are our bodies, ladies! It's OK to talk about them. And it's really, really important to get comfortable, get real, get knowledgeable and get healthy.

SEX AND THE LADY BODY

Our lady parts are functional, but they also put the fun in function. If you're sexually active, I'm hoping/assuming/insisting that you see an ob-gyn and take precautions that protect you from exposure to STDs and unwanted pregnancy, which can potentially impact your health and your life for the long term.

Here's another way to take care of your sexual health: exercise! Dr Chavkin explained to me that staying fit totally has a huge impact on our sex lives. As we get older, a weakened pelvic floor can put a damper on things in the bedroom. A weakened pelvic floor can have many causes, including obesity, vaginal childbirth and genetics, and is likely a combination of factors. It typically becomes an issue later in life, but some women in their late thirties and forties may also experience symptoms.

Exercising regularly and doing targeted pelvic-floor exercises like Kegels (see below) will help to keep your pelvic floor strong. Not only will keeping fit decrease the chance that you might experience pelvic-floor weakness later in life; it may also improve the quality and enjoyment of sexual intercourse and orgasms throughout your life. Who can argue with that?

There is a lot to be said about how sexy you feel when you discover your fitness. When you feel strong and healthy and comfortable in your body, you are likely to have more confidence and enjoyment everywhere you go, and that includes the bedroom, even if you aren't there to get any sleep.

HOW TO DO A KEGEL

Kegel exercises were developed for strengthening pelvic-floor muscles as well as strengthening the muscles that support the bladder, uterus and bowel.

Here's how to do a Kegel:

1. Identify the muscles in play. You can do this by lying down, inserting a finger into your vagina, and then squeezing. If you feel a tightening around your finger, then you have identified the muscles. (You can also find them while you are peeing: just try to clench and stop the flow in midstream.)

2. Once you have identified the muscles, practice with an empty bladder. Just:

   - Tighten the pelvic muscles and hold for ten seconds.
   - Then, relax your muscles completely for ten seconds.

You can do ten sets, three times a day, and you can do your Kegels while you are standing, sitting or lying down. But don't go overboard! Kegeling more than is recommended won't give you a vagina of steel, but it may cause muscle fatigue, which can make any underlying issues worse.

CHAPTER 22

THE ABC'S OF ZZZZ'S

———

NOW THAT YOU'VE POWERED yourself through a glorious day of sweating and pushing yourself and felt the rush of endorphins in your body, now that you've given yourself the nutrition to help your bones build and your muscles repair, it's time to do the most important thing you can do for your body: rest.

My sleep is very important to me. I don't keep any electronics in my bedroom. I wear a sleep mask. And I need a dark, quiet space. Sleep is when my body powers down so that it can restore, repair and replenish. When I get enough sleep, my memory is sharper. I can focus. And my body is more capable.

You know how it feels to be sleep-deprived – everything is fuzzy around the edges. It's easy to feel irritable and become inattentive. You might find yourself snacking because your body is searching for energy wherever it can find it.

If you don't get enough sleep for several nights in a row, your body will begin to produce more cortisol. Your jeans may get tighter, because cortisol encourages your body to store fat in your belly. You may get into more arguments. You may make more mistakes. Accidents may happen.

In fact, some of the biggest accidents in our history have been linked to lack of sleep. The nuclear accident in Chernobyl in 1986. The huge *Exxon Valdez* oil spill off the coast of Alaska in 1989. And recently, the tired banker who mistakenly transferred a couple hundred million pounds sterling instead of sixty because he fell asleep on his keyboard. Accidents happen every day because people are tired! According to the National Sleep Foundation, a hundred thousand car accidents each year are the direct result of drowsy drivers, and being tired may be the cause of a million crashes a year that are attributed to other causes. According to studies of medical students (a group of people who are often chronically sleep-deprived), residents who slept less than six hours a night reported making serious errors, coming to lousy decisions, consuming more alcohol, having more fights and gaining more weight than non-sleep-deprived people.

Mistakes? Bad decisions? Fighting?

And just think – you can avoid all of that if you just get enough delicious, warm, cosy, SLEEP.

Humans sleep . . . about one-third of our lives. If you are twenty-four years old, you have spent about eight years of your life asleep.

WHAT DOES IT REALLY MEAN TO SLEEP?

Every animal sleeps. And humans sleep a lot – about one-third of our lives. If you are twenty-four years old, you have spent about eight years of your life asleep. So, Sleeping Beauty, let's get down to it. Simply lying down and shutting your eyes does not count as sleeping. When you are truly asleep, your consciousness is reduced, your sensory activity pretty much halts and all of your voluntary muscles cease moving. Your involuntary muscles still perform their jobs – you can still breathe, of course, and your heart beats – but everything else stops. When you sleep, you are in an anabolic state and the growth part of your metabolic processes – and all of your systems – immune, nervous, skeletal and muscular, reap the benefits.

REM AND NON-REM

Yes, R.E.M. is an awesome band. It is also an acronym for rapid eye movement, a stage of sleep wherein your eyes move back and forth more quickly than they would if you were watching a tennis match, your brain activity is heightened, and you dream vividly. REM sleep makes up about 50 per cent of the sleep of children, but only 20 per cent of adult sleep. People still aren't sure exactly what REM sleep does for us, but some posit that REM is when our memory banks get a reboot so that we can keep thinking clearly as we age.

Most of sleep, however, is not REM sleep. Sleep tends to happen in stages and cycles; three stages of non-REM sleep are followed by one stage of REM sleep. Each night of sleep can have up to four cycles of non-REM and REM sleep. The first cycle is about an hour and a half long, while subsequent cycles can last up to two hours. During the deepest parts of non-REM sleep, the immune system is bolstered, tissues are repaired and bones and muscles are rebuilt. During the deepest parts of REM sleep, when we are dreaming, a slight paralysis of our muscles occurs to prevent us from acting out.

LIGHTS ... BRAIN ... INACTION

If you've ever been camping or spent time in an area without much electricity, you know that when the lights go off, it's time to go to sleep. Back when humans couldn't just flip on a light, when the sun went to bed, they burnt a candle or two and then followed suit. And they woke up when the lights came on again, that is, when the sun rose. Nowadays, though, especially if you live in an urban area, the lights are on all night long. And that's not just outside. Even indoors, even when you've turned off the lamps and the overheads, all of the technology that we have in our homes is still lit up, sending little beams of light out into the rooms where we are tossing and turning, wondering why it is that we can't fall asleep.

Our sleep cycles are regulated by our circadian rhythm, which is regulated by light and darkness. Are you getting the picture now? When it is light out, our bodies think we should be awake. When it's dark, our bodies want to go to sleep. There are two hormones involved in our sleep cycles, the same hormones that play a role in how our weight is gained and distributed: melatonin and cortisol.

Melatonin is designed for sleepytimes. Made in the brain by the pineal gland, then sent out into the bloodstream during darkness, melatonin encourages our peaceful slumber. Melatonin levels hit their max at about two am and hang out until four am, gradually ebbing so that we can wake up to greet the day.

Cortisol is the bright-eyed, bushy-tailed, wakey-wakey partner to melatonin. It is made by your adrenal glands, which can be found nestled right above your kidneys. Your cortisol levels will be lowest at night, when melatonin is watching the store, but will begin to rise early in the morning so that you'll be able to rise early too.

Back when humans couldn't just flip on a light, when the sun went to bed, they burnt a candle or two and then followed suit. . . . Nowadays, though, especially if you live in an urban area, the lights are on all night long.

As we discussed in Chapter 16, cortisol is also the hormone your body secretes when you get stressed. High stress levels mean high cortisol levels, so chronic stress impairs your ability to get a good night's sleep. And that can be stressful!

TECHNOLOGY AND MELATONIN

Our modern technologies present another reason for sleeplessness. All of our electronic devices emit blue light, a short-wavelength light that has been found to interfere with our melatonin production. Think about it: melatonin helps you sleep. And yet, your room is probably full of personal electronic devices, like tablets, phones, computers and televisions, all emitting these blue lights that block the very thing that helps us rest.

Do you usually use your electronics right up to the second you go to sleep? You may want to reconsider that habit. Suppressing the hormone that helps you sleep is *not* a great nighttime strategy. Lack of deep sleep has been found to contribute to a weaker immune system, as well as type 2 diabetes, obesity and heart disease.

So check your room out. How many glowing lights are there, like little green, blue and red eyes blinking at you from across the room and casting colourful shadows across the wardrobe? Shut them down before you go to sleep! Unplug, unplug, unplug. You can use a power strip and plug in all of your electronics in a neat little row, which will enable you to shut it all down with the flip of a switch.

And don't shrug your shoulders and then just forget that we talked about this! I know it sounds impossible in this day and age, but we have analog bodies trying to live digital lives. Biologically, we have not caught up to all this technology. So what is more important to you? Convenience? Or taking care of yourself, which includes your sleep patterns? So give it a try. For at least one week:

Stop using your electronics at least an hour before bed.

Turn off idling electronics or get them out of the room.

Sleep in a darkened room, blinds closed.

Then see how you sleep and how you feel in the morning.

CREATING A NIGHTTIME RITUAL

For the past twenty-five years, I have been on the move. I change time zones on a weekly basis. I sleep in hotel rooms more often than my own bedroom. And all of this moving around, which can be so disruptive to sleeping patterns, has taught me one very important thing: *nighttime rituals are essential.*

I might wake up one day in Los Angeles and then go to Sydney, where the local time is nineteen hours ahead of Los Angeles. I'll probably have to wake up by five or six am to start my day, which means I need to get my sleep so that I can do my work, which means that I need to be able to go to sleep, wherever I am, no matter what time it is in Los Angeles.

If I need to be in bed by ten pm in Sydney but it's morning in Los Angeles, it's hard for my body to understand that it's time to sleep now, not go for a run. So I have learnt to give my body a cue that it is time to go to sleep. I've created a ritual. And no matter where I am in the world, no matter what the room looks like or how comfy the pillows and bed are, I have a ritual that stays the same. I do it every night, and it cues my body and my mind to wind down, so that by the time I get into bed I can shut down everything completely and fall asleep soon after my head hits the pillow.

What are your rituals?

I'd recommend that after you have prepared yourself for the next day, by packing your meals ready to go and setting out your outfit or gym clothing or packing a gym bag, you find ways to get quiet with yourself and prepare your body for rest.

- **Close out the outside world.** Close the curtains or shades, turning off the TV, computer, phones.
- **Set your alarm.** Do it now so you don't forget, which will give you peace of mind as you sleep. I have friends who can't sleep through the night because they can't remember if they set the alarm or not, so they wake up in the middle of the night, worried that they've overslept, and then can't get back to sleep. So make setting your alarm part of the ritual and take it off your mind.
- **Prepare your bed.** Ideally, you make your bed in the mornings and it's a place for rest, not a place that holds all your dirty clothes or computer, so this should be easy. The bed should be a sanctuary, a safe haven, which serves only one or two purposes. You know what I'm talking about. Your bed is about sleep, slumber, restoration and sexy time. That's it!
- **Brush your teeth.** This is another part of the ritual that signifies the day is over. No brushing your teeth and then going to grab a snack in the kitchen. The whole purpose of cleaning your teeth is to make sure there isn't any food residue left. Bacteria breed like mad and you don't want sugar in there rotting your teeth. So *always* brush *and* floss your teeth before bed.
- **Wash your face.** Take that time to take care of your skin. Use a little face soap, pat it dry, moisturize. Washing your face is self-nurturing. It's time that you give to yourself to take care of yourself. A moment to look at yourself in the mirror and say, "Good job today! You worked hard, you did your best." Or, "Tomorrow we can pick it up a little, I have confidence

in you!" Acknowledge yourself. Check in with yourself. Let go of the day so you can close down once you climb into bed. Alternatively, you can take a hot shower or warm bath, though i.e. not everyone likes to bathe before bed, because for some it is part of a morning ritual.

I always take a shower before I get in bed, because I like to wash the day off my whole body. The sweat and grime from riding in taxis or travelling on a plane or sitting in the cinema, of hauling bags at the grocery store, picking up dog poop . . . when I climb into bed, I want it to just be me and mine in between the sheets, not the whole rest of the world. Showering is an essential part of my ritual – just a quick one-minute rinse-off with some hot water and soap to relax my muscles and make me feel like a wet noodle ready to be poured into bed. I also like to put a little moisturizer on before I go to bed, usually something with lavender, which helps me relax.

- **Then it's time to climb into bed.** Straight to bed! Turn all the lights off, don't check what's on TV or who is still up on Facebook (remember how short-wave lights affect your ability to produce melatonin).

MIRROR, MIRROR

When I say to have a look in the mirror at night and give yourself a "You go girl!" the point is not self-adulation: it's self-discovery. It can be about having a relationship with yourself that makes you accountable for the things you do and the person you are. A way of connecting to our interior selves as well as our exterior selves. Sometimes we imagine ourselves to be something that we're not – perhaps someone that we wish we were – but when we look into the mirror, we are faced with who we really are. And looking in the mirror doesn't have to be about vanity. Let the mirror be a friend instead of a judge. So often we look in the mirror and loathe what we see, but we should remember that the most important thing is to love what you are and be honest about who you are. Make time to stand in front of the mirror and identify all the different beautiful parts of your body that you love. Don't be shy! It's just you in the mirror, and you can love your body however you want to love it, without anybody else's validation or approval. The more you love your body and give it attention, the more connected you will be with it.

A little appreciation goes a long way. A lot of appreciation goes a really long way.

NOW, *SHHHHH*

Your sleep is as important to your health as your nutrition and fitness. Sleep helps your body conserve energy, because fewer calories are burnt while you are asleep. It gives your body time to repair muscles, create new tissue, build protein and release growth hormones – and some of these functions can happen ONLY while you sleep. So embrace sleep, the restoration period for your body and mind. Not only does it make you stronger, it helps you remember what you learnt today and be able to focus on what you want to accomplish tomorrow.

When you're asleep, new information that you learnt (like this chapter) is absorbed. Stresses are worked out through dreaming. And then you wake up refreshed, full of all the energy and confidence you need to conquer a brand new day.

PART THREE

MIND

You've Got This

CHAPTER 23

YOU'VE GOT THIS

CONGRATULATIONS! YOU'RE HERE. YOU'VE read through almost 200 pages of nutrition, biology, chemistry and anatomy. And I'm hoping that, like me, you have learnt a *lot* along the way. You know what hunger really is. You know the difference between whole foods and processed foods. You know what insulin is and what it does. You know how your heart and your lungs work together to deliver oxygen to all of your cells and how movement makes those processes more efficient. You know how your muscles and bones are affected by food and exercise. You know that your nutrition and your fitness work together to keep your body strong and healthy.

So now what?

Now you get to work. Now you turn information into action. Knowledge into practice. Wishes into reality. And you can do it. Because YOU'VE GOT THIS.

I say that because I believe it. I have seen girls who had no idea they were athletes become the fastest runners around. I have seen friends who could barely lob a tennis ball transform themselves into fitness buffs who could rock out in classes that totally kicked my ass. I have seen how strong I became when I started doing activities that built up my strength.

That is why I believe that individuals have power over their lives. That is why I believe that YOU have power over your life.

Now I need you to believe that. I need you to believe not only that you are capable of being a strong, healthy, vital, happy, capable, woman, but also that you deserve to be. That you are worth the effort and challenges it takes to become that person.

ADMIT THAT YOUR BODY IS AMAZING

Your body is amazing. It is. It is a powerful, incredible, intricate machine and by now I hope you've learnt enough about its inner workings to agree with me and that you're ready to start loving and respecting what you have instead of wishing for body parts that Mother Nature just didn't give you.

We all do it. We look at a friend or a girl passing by who has a physical attribute that we yearn for, maybe long legs, strong athletic arms, a flat stomach, curvy hips, narrow hips, a small booty or (if you're me) a big booty. You see how she fits into her clothing, how stylish and effortless it looks to be her, how lucky she is that she was born with the curves of a woman or the sinewy body of an athlete. We always want what we don't have! This is a *huge* trap that many of us get tangled up in. When you're caught up in this trap, you tend to ignore the needs of your own body because you're too busy hating it for what it ISN'T instead of loving it for what it IS.

And who wants to live in a trap? Not me. Not you. So let's set ourselves free. Let's become aware, become conscious, get mindful. What's the point of walking through life totally oblivious to what you're doing and how those choices are determining your mental, physical and emotional health? That SUCKS! You can't get the most out of life if you aren't aware of how to get it. And it only gets worse over time. As you age, your body starts to react more and more to misuse and lack of proper care and maintenance. Trust me, I know that it only gets harder.

No matter how old you are, now is the right time for you to start connecting to your body and making the choices that will allow you to age gracefully into your hundreds: healthier, stronger, sexier, more capable, more creative, more interesting. You have so much life to live, and I want you to live it to your fullest, in the strongest, most healthy body that you can inhabit.

EMBRACE THE IDEA OF GROWING OLDER GRACEFULLY

A lot of us fear the idea of ageing – getting older, being less mobile, having less energy and being less of ourselves. But I look at ageing like this: getting older is a *blessing* and a *privilege*, and if you lay the foundation for a healthy life in your younger years, your older years may very well be some of the best of your life.

And the point I'm making is not about keeping up your youthful appearance. This is not about beauty and the aesthetics of our bodies. I want you to *feel* young. I want you to *feel* strong.

When I was a child, I always loved being with my elders. I was obsessed with my *abuelos*, my father's parents, and my grandma, my mother's mother. They were endlessly fascinating. They did things effortlessly that seemed impossible to me. My grandma, until she was about seventy-five, raised all of her own livestock and grew all her own vegetables in her backyard in the valley just above North Hollywood. She would carry over twenty kilograms of feed for her chickens, rabbits and goats over three kilometres in the middle of the summer heat. And I swear, my *abuelo* could fix anything with just a paper clip, some duct tape and a metre of rope. They were my superheroes, and I wanted to know everything that they knew.

I also loved the way they looked. Their skin was beautiful to me: the wrinkles told the story of their lives, the joy and the sorrow, the hard work that their bodies had done for them their whole lives. The strength in their muscles that they still possessed at an age when their bodies could have become weaker but instead were utilized with skill and purpose.

These were my role models for ageing, for growing older with strength and ability. It never occurred to me that it was a bad thing to grow older, and now I live in a world and work in a business that is bent on telling people – especially women – that they are no longer vital once they start to show signs of 'ageing'. It makes me sick to my stomach! I am horrified by how deeply these ideas have permeated our culture, and I worry about the young women who are being influenced by this nonsense.

It breaks my heart and frustrates me no end that our society values youth over experience. How silly is that when it is physically impossible to stay young? And when experience gives us wisdom we could not possibly have had as kids? Our bodies age every single day that we are lucky enough to be alive. The alternative to not ageing is as grim as it gets, because if you're not getting older, you are dead. Taking good care of yourself is a wonderful way to slow down the rate at which your body ages, because at some point, how old your body feels is not a question of years, but a tally of habits and choices and chance. But no matter how much we exercise and how much we moisturize, it is the law of nature and the journey of being human that our bodies are ageing and changing every single day.

Instead of obsessing over staying young forever, isn't it better to want things that we *can* have, to aim our energies at achieving results that are actually achievable? The truth is that the best possible result of our fitness and nutrition work is that we can age gracefully, by which I mean healthfully. I for one can say honestly that I feel better and stronger and more capable now than I did when I was twenty years old, because I've taken better care of myself in the last fifteen years than I did in the first twenty-six years of my life.

Ageing healthfully is ageing happily.

I am so grateful that I got to make this discovery, and I hope that I can convince you that it is worth your while to embrace this idea too. Being healthy is your freedom, your independence, your ability to learn new things and spend time with friends and family and climb a tree if you feel like it. The more you learn the best ways to care for yourself and apply that information consistently, the better you will feel and the more you will be able to really LIVE, not just today but for the rest of your life.

Who wants to be young forever? I'd rather get to live long and thrive throughout my life. And I want to live in a body that I love and respect because it is the body that gives me the ability to lead that life. As I write this book, I am looking forward to my forty-first birthday. I'm happy to state my age because I believe that getting older is the best thing that has ever happened to me. The knowledge and wisdom that accompany age can make life easier and more joyful.

Getting older is *amazing*. It's what life is all about, even if it feels strange sometimes to see the body that you live in change. If you do it right – and I mean this, because you really have to do it right – if you take responsibility for yourself, and really do the work, you will be able to *love* getting older.

TURN KNOWLEDGE INTO ACTION

There is no magic potion for health. There is no trick, no pill, no incantation. But there is knowledge, and there is action. And information without action is just a set of facts. If you want health, you must turn information into part of your daily routine, or else all of the time that you invested in reading this book is useless. You might as well use it as a doorstop now.

Information is power only if you USE it. Only if you practice it.

I mean, think about it. Think about all the doctors and health professionals that you've met. Are they all healthy? Are they all fit? Chances are that some of the people advocating that you live a healthier lifestyle don't put that information into practice themselves. How is that possible? It's possible because having knowledge and taking action are not the same thing.

You want it? Then you gotta *do it*. Let me say that another way. *You* have to do it. You have to commit to learning the basics and turning knowledge into action. Like reading up on your biological basics, the way you have been in this book. Eating more green salads. Buying vegetables at the market and then looking in books or online to learn how to prepare them. Choosing whole grains instead of the processed versions. Drinking water throughout the day. Moving consistently, all day long.

Health is not about depriving yourself. It's about giving yourself everything that you deserve. And that begins with being kind to yourself, and gentle, because everything takes time – but if you persevere, you will get there. Think about how long you have been living the way that you have. How many years of the same habits, the same attitudes, the same beliefs? My guess is that it's been a pretty good run of all of those things. If it took you that long to master those habits, thoughts and beliefs, then I would like to ask you to be patient with yourself while you learn to take different actions and form new habits and beliefs. Nothing happens overnight! Do you hear me? You cannot expect to make a change all of a sudden and have it stick without any hiccups or challenges. Life is about determination and practice. And the only way that you really get good at anything in life is by doing it over and over again and doing it CONSISTENTLY.

"Wait," I hear you saying, "but that sounds hard."

Well, I didn't say it would be easy. Any goal you've ever achieved, any accomplishment you've ever had, probably took some serious dedication and work. You had to show up every day and put in the time and put in the energy. Same goes for your health.

It might not be easy at first, but I promise that it will get easier. At the beginning, making new choices is always a little bit uncomfortable and scary, particularly if you're challenging deeply ingrained habits. That's why I don't want you to think of this as a diet, a lifestyle change, or anything like that. It's a *learning practice*. If you are scared because you don't want to fail, please remember that this is not something you can get right or wrong; it's just something that you do your best at, again and again. And eventually, it's just something that you do. It becomes A WAY OF LIFE.

STEP 4:

STRIVE FOR CONSISTENCY

The key to longevity and good health is *consistency*. If you consistently make bad, unhealthy choices, then you will likely be unhealthy. If you consistently make good, healthy choices, then you will likely be healthy. *Whatever you are consistent in is what you will become.*

Your health is an equation. If you have ten opportunities to make a choice and you choose unhealthy eight out of the ten times, then your consistency is shot and you're building negative habits and an unhealthy body. If you choose healthy eight out of ten times, you're building positive habits, strengthening your ability to continue to make positive choices and creating a healthy body.

So that's where you start. Choice by choice. Slowly improving on where you've been. If you *always* make the unhealthy choice, even balancing it out to fifty-fifty creates an improvement. When healthy becomes the dominant choice, you will feel the effects accordingly.

Here's a good way to test what I'm saying: pick one thing that you think you can make healthful choices about consistently. Maybe it's giving up fizzy drinks and juices for a week. If that's the case, then every time you want to drink a fizzy drink, choose water instead. That means that instead of filling up with a 1-litre bottle of a fizzy drink from the petrol station, you buy one of the 1.5-litre bottles of water.

Do that consistently for a week. Even if you decide to have a fizzy drink, choose a smaller one and make water your main beverage regardless.

Just see how you feel at the end of that week. See how the taste of fizzy drink shifts when you let your mouth get used to *not* being doused with sugar all the time. See what changes you feel in your body, in your energy. And then do it for another week, and see what it feels like at the end of that week. If you don't like it, you can always go back to your fizzy drinks, right?

If this new habit feels good to you, you can start adding other healthy choices, like swapping the cheeseburger for a piece of grilled chicken, trading potato salad for a green salad or having an apple instead of an apple pie.

The more consistent you are with these choices, the closer you come to being the happy, healthy, beautiful, glowing woman that you want to be. Being healthy and well is a lifelong endeavour. It's important to establish a relationship with your body, to understand what its needs are and how to fulfil those needs. To connect to how your choices affect you physically, mentally and emotionally.

Our human selves are so delicate and yet so resilient, and understanding what we really need makes all the difference to our experiences in life.

STEP 5:

AIM FOR THE YEAH! FEELING

Have you ever had the feeling that everything is just as it should be? Where you feel relaxed, energized, content and confident? Where it seems like everything is going to go your way, and if it doesn't, you'll be able to deal with it? It's the kind of feeling that says, *YEAH!! I can do whatever I want to do!*

When I'm paying attention, I notice that the YEAH! feeling is often the result not of an event that occurs outside of me, like getting something that I wanted, but of something that happened *inside* of me when I gave my body what it needs to thrive. When I'm really using my discipline to give myself everything I need to feel my best – wholesome foods, a lot of movement, fresh water, enough sleep – that's when I feel like, YEAH! It's truly the best feeling in the world! When I'm in that place, I understand the real meaning of joy and happiness. Everything is just easier, even all the choices it takes to maintain that way of life.

But there's also another very human feeling that's basically the exact opposite of "You go, girl!" It's more like, "Why try?" It's that "What am I doing?" feeling. Like you can't do anything right. Like you shouldn't bother anyway. Or maybe for you it's "Why can't I make better choices for myself?" Or "Why can't I follow through on my best ideas?" It's a feeling of defeat.

And when I'm being aware and paying attention, I notice that that "Everything is wrong" feeling generally crops up when I'm not taking care of myself the way I should be – when I'm not paying attention to my nutrition, when I've let my fitness slide, when I'm too stressed out or when I'm overly tired because I haven't gotten enough sleep. Unfortunately in today's world, it can be easy to slide into that way of life. For me, that becomes especially true when I'm filming a movie.

When I'm on set and have to wake up at five am to work a twelve-hour day or sometimes longer, it can be a challenge to find the time to go to the gym and train. If I don't plan what I'm going to eat for the day and take it with me, then I'll end up eating food I don't want to eat, just because that's what's available. I can fall into the rut of not sleeping enough, because when I finally finish work for the day, either I'm too wound up to go to bed or I want to spend some time with my friends and family.

If I find myself in a pattern of poor nutrition, poor sleep and not enough physical activity, I notice that it comes along with a curious feeling . . . everything from my thoughts to my actions to my emotions starts to take a lot more effort than when I'm taking care of myself properly. When things get too busy or I'm not being conscious enough, I start to feel the drag, and the longer I wait to get back into my best habits, the more I can feel that difference taking hold. I'm more impatient. My emotional heart feels tired, my spiritual mind feels worn down and my physical body feels exhausted.

And it's that kind of exhaustion that makes us weak and susceptible to illness.

But since I've learnt that my mental and emotional health are directly connected to how I care for my body, as soon as I start to feel that drag, I get myself back on track. I make sure to get to the gym, no matter how early or how late I have to go. I make sure I have the food I need to support my energy throughout the day. I make sure I get eight hours of sleep.

And then? There's a spring in my step, and an extra bit of patience for a challenging conversation, and best of all, the simple pleasure of feeling *goooood* all the way through: mind, body and heart.

IN SUMMATION . . .

Even though it isn't easy to make changes, it is possible. And by 'isn't easy', I mean, 'can be really super-freaking hard'. Human beings love to get set in our ways, but if we want to change, if we open ourselves to learning new things and work to create new habits and patterns, we may discover that we are wonderful cooks even though we thought we couldn't fry an egg, or that we can do ten chin-ups even though we thought we couldn't do one, and many, many other things that will be unique to your individual journey.

It starts with information and developing self-awareness, and it continues with commitment, dedication, and action. And the confidence to be able to say, "I've got this."

You can do this. I know it. That's why I spent years talking to experts and gathering all of this information for you and for all of the people who ask me questions about what to eat and how to eat and what kind of training might work best for them, because we all deserve to have this information. This information doesn't just belong to GPs and personal trainers and people with gym memberships or personal nutritionists. It belongs to YOU.

So please, make it your responsibility to keep your body strong and healthy, because no one can do it for you. *You* have to be the one who puts in the time and effort. Even if you have someone helping out and offering advice, *you* are the one who needs to follow through. And *you* are the one who will benefit from applying that knowledge and putting it into action.

CHAPTER 24

THE STATE OF BEING CONNECTED

J UST NOW, I GOOGLED 'mind-body connection', and got more than forty million results, gathered everywhere from the Mayo Clinic to Harvard to the National Institutes of Health. That's a lot of people in a lot of places talking about the mind-body connection and how it affects our lives and our health. So what does that have to do with you? A lot. Because your mind-body connection is a crucial part of the self-awareness that helps you turn knowledge into action.

Connection is a word I use a lot, because everything is connected! As we've learnt, the chemical and hormonal processes in your body, like whether or not you get your period or how well you sleep or where your body stores energy as fat, are directly tied to your diet and physical activity level. And those chemicals and hormones also have an effect on your brain . . . because they are all part of the same package of *you*.

You have a thinking self and a physical self and an emotional self; when I use the phrase 'mind-body connection', I'm referring to the ways in which all of these selves interconnect. Your body is influenced by your mind, and your mind is influenced by the chemical and hormonal processes in your body, and your emotions are influenced by the way you care for your body and your overall health. That's the power of connection!

If you are taking care of your physical body, your mind will benefit. And as your mind benefits, you will find yourself more equipped to take care of your physical body. It's kind of crazy if you think about it, that it's just this simple: how you treat yourself is directly connected to how you feel. The choices you make as to what you eat and drink, and whether you get enough sleep, and how you move are all connected to what kind of a day you have – and ultimately what kind of a life you have.

Connections!

You may not understand what I mean right now, but if you trust me a little, if you try, it will come. If you learn to slow down and focus, to relax and think about how your body feels, to *communicate with yourself* so that you become aware of your body in a consistent way, it will come. The key to achieving that state is allowing your mind and your body to communicate and freely exchange information . . . information that you can interpret to change behaviours that are making you feel worse and reinforce behaviours that are making you feel healthier.

By becoming more aware of the relationship between what you do and how you feel, you can begin to make subtle shifts that will encourage more positive shifts. By waking up to the connection that already exists between your mind and your body, by strengthening that connection, you will gain understanding about how your food and movement patterns relate to your moods, your energy levels, your entire life.

WHAT IS YOUR BODY TRYING TO TELL YOU?

Have you ever had a conversation with someone who wasn't paying attention to you? Perhaps you raised your voice, or tugged at their sleeve, or waved your arms over your head just to get them to *listen*. We all want to be heard when we have something important to communicate, and your body is the same.

Many people try so hard to ignore the way they feel, in their bodies and in their minds, that even when their bodies and minds are SCREAMING at them, in the form of illness or anxiety or weight gain or depression, they can't hear. Connecting MIND and BODY means learning to listen, learning to understand the messages your body is sending you and giving your body what it needs.

And if you don't listen to its most gentle messages, they're just going to get louder. Why? Because your body wants to survive. Your body wants to stay alive, and the only way for it to do that is to let you know when you're doing something to HURT yourself instead of HELP yourself. These messages are not casual texts; they are major transmissions.

Your body has developed lots of warning signs to alert you to when something isn't right. Heartburn and indigestion, for example, are your body's ways of telling you that you've eaten something that it can't digest properly. What usually happens when you get indigestion? Do you take an antacid and forget about it once the discomfort has passed? Or do you backtrack and try to recall what you've eaten that might have caused that reaction, and make a note to avoid eating that food the next time you're offered it?

When you have a headache in the middle of the day, do you swallow some ibuprofen and wash it down with some coffee or diet soft drink? Or do you ask yourself if you might be dehydrated and consider how much sleep you got the night before?

Our bodies speak to us all day long, from our hunger to our sleepiness, through signs that can be subtle or overt. These signs have been carefully designed by the body to tell you exactly what it needs; it is not a mistake or a fluke, but a true language that is real and very clear if we listen. For instance, consider a yawn. A subtle sign like yawning can mean that the person sitting across from you just yawned (it's contagious!), or more likely it can mean that you are dehydrated and need to drink some water or that you're running low on fuel and need to check in with your hunger level, or that it's time to get up and take a walk around the office or go outside to get some oxygen and get your blood flowing.

When you don't pay attention to these small warnings, you inch closer to the giant SOS signals: illness, like diabetes, high blood pressure and obesity. So listen up, and give your body what it's asking for, the first time it asks.

THE MIND-BODY CONNECTION
WORKS BOTH WAYS

A body is a physical entity. So is a brain. The mind is something wholly other. It can be felt but it cannot be touched. It holds your dreams but it cannot be held. Yet it still speaks to us, and it is your job to learn how to listen.

Part of the mind-body connection is the realization that sometimes when your body hurts, the pain is a signal from your mind telling you that it's hurting emotionally – symptoms like aches and pains, headaches or fatigue can all be signs of sadness, stress or depression.

We've all heard the term 'body language', as in reading people's emotions or intentions from the way they hold themselves. We might be able to tell if a friend is in pain by the way his shoulders are hunched over, or that your boss is angry because her arms are crossed, or that a guy likes you because you always find him standing (awkwardly) close to you. Our bodies send messages to the people around us, and they also send messages to us.

A lot of people feel things physically before they recognize them emotionally. Or what feel like emotional needs may really be physical needs. When your mind tells you something like "I want a hamburger", there's a chance your body is craving protein and iron. Or when you're at work and your mind says, "I hate my job and I suck at it", it may be that your body needs a break and you need to get some fresh air or move or sleep.

If you want to wake up to the connection between your body and mind, you must check in with both on a regular basis. What I mean by 'checking in with your body' is just that: actually spending quiet time listening to what it is trying to tell you by paying attention to the way it feels. Because your body does speak to you; you just have to learn to listen. And guess what? When you start to quiet down and listen to your body, your mind will start to speak up too. It will have reactions to the way your body feels (*my knee has been hurting; maybe I should make an appointment to see my GP*), and some concerns of its own (*I'm so tired lately. How have I been sleeping?*).

Some people like to meditate or do yoga; others have daily stretches or a regular morning hike where they pay attention to how they are feeling. These rituals are so important because they give us a set time to check in with ourselves. We all need rituals like these.

CONNECTION EXERCISE

Here's a little exercise that you can do to start connecting to your body. It's a subtle and quiet way to tune in to your body one-on-one. You can do it as a part of your morning or evening ritual or whenever you can fit it into your day.

STEP 1: MAKE SPACE.
Pick a time when you can relax. Find a spot where you can lie down and spread out. Maybe it's on your bed or on your couch or on your sitting room or bedroom floor.

STEP 2: RELAX.
Lie down on your back, and let your body sprawl. Maybe your body wants you to spread your arms and legs like a snow angel, or maybe you want to fold your hands across your chest with your legs bent.

Pay attention to how your body feels. Is it comfortable? What does it want you to do? Whatever it is, do it. If you can't quite get comfortable, take note of that, too.

STEP 3: BREATHE.
Once you're in a comfortable position, bring your attention to your breath. You might take a deep breath in through your nose and blow it out through your mouth, really pushing it out at first, then start to relax into your breathing naturally.

STEP 4: MOVE.
As you relax into breathing, move your body. Turn your legs side to side, twisting at your waist. What does that feel like in your hips, your waist, your legs? Now move your arms, maybe to support your movement of your lower body, or maybe they feel nice resting across your forehead. Then stop thinking about telling your body how to move and start just letting it move on its own. As you take each breath allow your body to find its movement. Nothing it does is wrong!

STEP 5: LISTEN.
If you keep breathing and moving at the same time you will find that your body is speaking a language of movement to you. It is telling you how it needs to move, whether that means sitting up to touch your toes or crawling onto all fours and arching your back.

The more you get used to listening to how your body feels, the more you'll find that every movement you make throughout the day is an opportunity to connect with your body. You'll connect to hamstrings in the back of your legs and your glutes every time you step up onto a stair. You'll start to see how your stomach tightens up when you go to pick up your purse and how your biceps curls it up onto your shoulder or forearm.

Move mindfully instead of mindlessly, and you will open a continuous dialogue with your body.

When I want to listen to my body and to my mind, I begin to stretch and allow my body to flow with movement and breath, breathing in and out, reaching in whatever direction it feels like going. When you begin to move and stretch and listen, your body will start teaching you its own language, and now all you have to do is note which parts of it are screaming at you and which parts of it whisper quietly.

This is the important part – the listening and the noting. When I want to connect, I think about how I feel all over, from the muscles in my neck to the muscles in my back, down my legs, even in my toes. I think about how I feel emotionally: do I feel sensitive, relaxed, buoyant? Every direction that I move in, I shift into the feeling of the stretch . . . sometimes it hurts . . . sometimes it gives me a huge sense of relief . . . I ask myself things like: Do my muscles feel tight or stiff or strong or weak? How easy is it to touch my toes? Does the right side of my body feel amazing when I reach over the top of my head and does it make me want to keep reaching even farther? Sometimes the way I move surprises me, because my body begins to lead my mind as much as my mind usually moves my body. With every stretch, I go a little farther. I connect a little deeper. And I learn about how I feel right then at that moment, physically and mentally and emotionally, and that helps me go forwards into my day.

NO MORE MIND GAMES

If you're anything like me, your mind has voices that tell you things all of the time, not only when you're checking in. It may be desires or insecurities, hopes or fears. And they don't live only in your brain. Sometimes they feel like thoughts, but at other times they may seem to emerge from your gut or your heart; you may identify them as ideas or needs or urges or emotions or cravings.

Like I've said, the only way to figure out what is going on is to start listening. Sometimes the voices totally contradict each other, like when you have the urge to take a chance but be careful, at the same time. They can be protective, like the voice that reminds you of your last broken heart and tells you to be careful about a new guy. They can be fearful, like the voice that tells you that maybe you should pass on that invitation to go to the beach with your friends because you don't want anyone to see you in a bikini. They can be helpful, like the voice that reminds you to slow down when you're driving because the last time you approached that hairpin turn in the road, you took it too fast and nearly crossed into oncoming traffic.

These voices crop up with guidance – good and bad – on a daily basis. They can appear anytime – when you're working or working out, when you're loving or being loved, when you're planning your meals or what you'll wear tomorrow.

DISARMING THE DOUBTING MIND

Some of the voices in our heads don't really have our best interests at heart; they want to sabotage our success. They are the ones that say, *You can't do it. You'll look stupid. That piece of cake will make you feel better.*

That voice likes to try to convince you that you are not capable of doing something that you are actually perfectly capable of doing. For instance, have you ever watched a marathon and thought, as the runners went by, *I could never do that*? Or stared at a packet of biscuits and thought, *I know these are going to give me a stomachache, but I've had a bad day so I might as well have a few treats*?

We've all had those moments when our mind somehow gets us to do something that's not in our best interests. There have been way too many nights when I wanted to be in bed by a certain hour so I could get my sleep and be able to wake up early before work and get to the gym, but I got distracted and busy with something in the evening and then found myself up way later than I wanted. And of course in the interest of getting sleep, I can talk myself into skipping my morning workout.

How is this possible? How can we know something on the one hand, how can we want something – and then go ahead and do something else that we know we shouldn't, that moves us further away from what we want? Every one of us has had the experience of thinking: I'm going to do Thing A, definitely Thing A, and then the next thing we know, we are doing Thing B even though it is the OPPOSITE of what we thought we wanted to do! It's *so* easy to do.

That's the part of the mind that is controlled by fear. It offers up a negative view of itself; it whispers to us that we are incapable, just because it's afraid. So it tells you that you're gonna eat some biscuits anyhow, so why not just get to it? And why eat just one when you know you don't have any will-power? Just eat them all! Then after you've eaten the biscuits, that voice says, *See? I knew you were going to eat them all. You'll never be able to take charge of your health.*

Where is this fear coming from?

Doing new things can be scary. There is great comfort in the familiar, even if the familiar is what is hurting us. Fear of discomfort, uncertainty and failure is just too much for the frightened, weaker part of our mind to bear; it doesn't know how to manage those feelings and is afraid that it will feel that way forever. So it panics and just shuts down any possibility of experiencing those feelings. Its goal is to keep you in its own comfort zone, the place that is familiar to it, where it knows what it will feel even if the feeling is something that is painful and defeating to you, like eating a whole packet of biscuits, which makes you feel physically sick to your stomach and totally emotionally defeated because you know what that packet of biscuits means to your health.

To your mind, at least those feelings of sickness and disappointment surrounding that packet of biscuits are familiar emotions that it deals with all the time – pain that is recognizable is easier for it to deal with than an unknown pain. This is how your fearful mind stays in control, by allowing you to defeat yourself.

LISTENING TO THE KINDER MIND

There's also a kinder mind tucked away in there, and that mind has a different message. It says, *I know this is hard but you can do it.* It says, *You'll feel better tomorrow if you don't do this.* Or, *You'll feel better later if you do this now.* These voices want the best for you, like the ones that say, *Pack your gym bag and take it with you! Eat a kale salad for lunch!* We hear them, but we often ignore them because it's hard to hear the truth sometimes, and this voice speaks only the truth.

This is the voice we have to tap into and listen to carefully. Because our minds are powerful – so you must remember that the same power that can help you to fail can also help you to succeed. Anytime you start to feel that your mind is pushing you in a direction you're not on board with, remember that there's another saying you've heard a thousand times (don't roll your eyes!): Mind over matter.

Because sometimes, even when you're certain of failure, you try your hardest anyway and actually succeed. Sometimes you allow yourself to think *I can do that*, and then you actually *do that*. That's mind over matter. That's using your mind instead of letting it use you. That's the basis of all human achievement. It's using the part of your mind that is the small voice inside you, the quiet steady one coming from your gut that's always there but can be hard to hear because the self-sabotaging voices always seem to drown it out. The smaller voice is the one that tells you the truth; it is the reasonable one that truly believes in you and knows what you are capable of. It's the one that's always rooting for you. It knows the difference between right and wrong and how much discomfort and pain you can endure to achieve the goals that you want in every living moment. This voice will always be there for you when you need good advice or when you need someone to push you and motivate you.

Not only do we need to listen to that voice, but we must honour it. That means that you need to do what it gently tells you to do and knows you are capable of doing. And once you begin to do that, you will find that your happiness will grow.

Because that voice speaks from truth. From kindness. From self-acceptance. Not from ego. Not from fear. I hear it, but, more important, I *feel* it. It's the voice that comes from deep within me, and it knows what is best for me in any given situation. We all have a gut instinct. Anytime you've been faced with a major decision in life, people have probably said, "Follow your heart" or "Listen to your gut." They're talking about tuning in to the guiding voice that speaks to you by giving you a feeling in your body. It's a visceral, instinctual, intuitive feeling. It knows you better than anyone else because it *is* you.

Here's why listening to your gut is so important: Happiness comes from within. It is the absolute truth. You may have to hear it a million more times before you either believe it or truly understand it. *But the whole reason that I'm writing this book is so that you can hopefully move a little closer to it and ultimately achieve it by applying all of this knowledge to your life. When we connect our minds and our bodies, we are our own best allies.* When we follow our inner wisdom, we are our own best mentors. That goodness on the inside spills out onto the outside. I really want for you to know what it feels like to be truly happy. Being connected to your body and your mind, and having them be connected to each other, is a major chunk in the puzzle of being happy in life and living it to the fullest.

TRUSTING YOUR GUT

For me, indications of how I really feel about things come from my gut. It's a subtle feeling in my stomach when I consider one option versus another. It feels kind of like I'm being kicked ever so gently from the inside of my stomach; at the same time, it feels like my stomach is dropping – subtle but undeniable. Or there's a quiet, steady feeling, like everything is OK and I'm safe.

My gut feelings give me guidance. After I weigh out the logical and logistical information that I have about any given situation, after I let my brain chew on all of the available information, I let my gut have a turn.

Let's say I am invited to do a film. I will think about the script and how much I love it. I will think about how, if I'm shooting on location for three months, I'll really miss my home and my family. Or it may be a challenging role, and I will wonder if I'm really equipped for it. First I will think about the pros and cons that relate to the facts. But then I will dig down deep inside myself. Am I up for the challenge? Is it worth it to leave my home behind?

Ultimately, the deciding factor will be the feeling that I get in my gut. Either I'll experience a sensation that this is the right thing to do, or it will tell me that I should pass and direct my energy elsewhere.

There are two ways that I achieve this. If I have some quiet time, I'll sit down somewhere peaceful and try to clear my mind of all the thoughts running around in my head. I just sit and breathe and listen to what my gut is telling me. The answer might feel like a single word, yes or no. Or it might be a lengthier explanation. Either way, my gut always tells me the truth. But often, I have to make decisions from a busy, noisy space, where I don't have a moment to sit and be quiet. Then I try to notice, while I'm processing the information, how my gut is reacting. Is it tightening up and miserable? Or relaxed and happy about what I'm hearing?

And then, before responding, I take one last check-in, and I always go with what it tells me. YES, this is a good idea! Or NO, back off, this is gonna lead you down the wrong path.

You can never go wrong if you listen to your gut! Even if stories turn out to have a different ending than you imagined, remember that when you made the decision, you were doing what you thought was best at the moment.

That is called honouring yourself, and it's one of the most important things you can learn to do.

WHAT DO YOU REALLY WANT?

So now that you're listening – what do you hear? What does your inner voice want? As you ponder this, think about the fact that the ability to consider what you might want is a huge gift. Being able to apply rational thought to our feelings and emotions – to consider how we really feel, deep down, not just on the surface – is what makes us human.

For me, I know that the more I listen to the inner voice that says, *You can do this*, and the more I trust the voice that tells me that I am worth the effort, the more I believe in myself. And that belief gives me the strength to trust myself – my true self, my confident self, my healthy self – and make choices that support my growth, my learning, my health, my evolution.

When you take a step back and become aware of how you see yourself, what your choices are, how those choices affect you and what you want in life – when you do that, it's like becoming conscious. It's like waking up. There are people who spend their whole lives avoiding that consciousness, living in a dream. All of their decisions are reactions to what happens to them, and there's no planning or goal setting or forethought. Their life is like being a pinball in a machine, being jolted this way and that by every stimulus that crosses their path – and they wonder why it seems they can never be happy.

Wouldn't you rather wake up? Wouldn't you rather see the stimulus, understand the reaction and choose for yourself what you would like to do? The key to health and happiness is knowing how your body works and knowing how *you* work – having that relationship with *yourself*. Knowing yourself allows you to listen to your inner voice and make decisions that are good for you.

So let's start to understand one of our most basic human reactions: the way we react when we encounter lots and lots of available food.

CHAPTER 25

DECODING THE MODERN APPETITE

MODERN HUMANS ARE EATING themselves to death. We've already talked about how crazy that is, since food is meant to help us LIVE. We've talked about how our bodies use hunger signals as a way of letting us know that they need nourishment and energy; we've also talked about how many of the diseases we're always hearing about as 'epidemics' are ones that we are bringing upon ourselves to a large degree. And we've talked about the best ways to fuel and nourish our bodies so that we feel good, look good and are happy.

It seems so simple: Eat whole foods when you are hungry. Stop eating when you are full.

So why don't we?

We've all been to a holiday gathering, sitting around a table laden with delicious food, and eaten so much that at the end of the meal we have to collapse on the couch with our jeans unbuttoned. We've all sat in front of a movie with a huge bowl of popcorn and wondered at the end where it all went.

So why are humans – who are smart enough to learn how to fly, to learn how to get to the moon, to invent the Internet – foolish enough to keep eating food until we make ourselves sick? And I don't just mean sick with diabetes and other diseases, I mean sick at the moment, literally nauseated because we have stuffed our bodies beyond our need for immediate nourishment.

WHY DO WE DO THAT?

The answer lies in our genes. And in the word APPETITE.

A couple of definitions:

HUNGER, as you learnt in the first part of this book, is your body's cue that you need nutrients. It is a biological need triggered from within us.

APPETITE is totally different. Appetite is the desire to eat because the sight, smell or taste of food makes you *want* to eat even though your body is nourished. Our appetite is roused by forces outside of us.

APPETITE FOR SURVIVAL . . .

Back in the days when we had to scavenge for our food, we could never be sure when our next meal was coming, so we would eat whatever was in front of us, whenever it was available. Back then, our ability to overeat was a *blessing* because without that ability to be gluttonous in the short term, we might not have had enough energy to make it to our next meal. And so as the body was genetically wired to store fuel as fat on the body, the mind was wired to find that food, recognizing it through sight, sound, smell and feel.

As we developed agriculture, we had more control over our food – we could produce enough excess to store for longer periods of time and rely on that food to get us through the winter. But for the most part, we would eat seasonally, and we would eat only the foods that were available to us in whatever part of the world that we lived.

. . . APPETITE FOR DESTRUCTION

One of the most drastic shifts in our world is how accessible food has become because we have so many methods of storing and preserving it. And that accessibility works against us. Food is no longer available only seasonally or when we track it down and kill it. It is *always* there. And so it is essential to understand how hunger and appetite work if we want to stay healthy in the face of this overabundance of processed, imported, convenient, inexpensive foods.

Hunger is an innate bodily need. But *appetite* can be triggered by any number of common cues. A cue may be the visual stimulation from a billboard or the sound of a TV commercial, the smell of burgers on a charcoal grill or walking past a favourite restaurant or a bakery (which reminds you of the delicious cupcake you had the last time you were there). And cues, those little reminders that we might want or need something, are usually followed by a want or a need for that thing. So even if you're not hungry, your appetite will suddenly be prompted, and you'll find yourself feeling like you have room for some more food. And that response is directly related to our inner cavewoman,

When we don't understand why we're eating or even where the urge is coming from, it's easy to become confused about our food choices.

who found food through her sensory perception and relied on memories of food to find it again. All of these modern-day cues speak to the genetic code that is STILL ACTIVE INSIDE OF YOU, that was once there to ensure your survival but now tells you to stop for a doughnut on the way to work after you've already eaten breakfast at home.

When we don't understand why we're eating or even where the urge is coming from, it's easy to become confused about our food choices. That's where biology comes in.

THE HIJACKED BRAIN

We've already discussed that your brain is a greedy organ, sucking up more than 20 per cent of the energy you eat to keep all systems running smoothly. So your food choices power your brain. But your brain also powers your food choices, acting as a prompter that tells you to eat certain things, avoid others, and identifies sugar as a potent, desirable DRUG. Yup, a drug.

Here's how it works. Think of eating as a chemical reaction from beginning to end, because it is. You already know how your body breaks down food into basic nutritional components, and that your hunger is prompted by your body's need for nutrition. When you are hungry and you eat whole foods, your body's needs are satisfied. But when you eat foods that include a lot of added sugars, those sugars bypass your 'I'm full' system and keep your brain asking for more.

You know that added sugar isn't good for you. But have you ever thought about why it *feels* so good? I mean, come on, we've all had that feeling! When you're eating something that tastes SO GOOD that even though you know you've had enough, you just can't stop yourself from having *a little bit more*? Maybe it's that piece of caramel-drizzled cake or that perfectly chewy biscuit,

Your brain loves endorphins, so even though you are out of control, overeating, and probably freaking out, you feel great (at least for a moment).

and you just can't stop thinking about it until you've eaten the WHOLE THING. Well, there's a really good explanation for that. And it has nothing to do with how 'good' or 'bad' you are.

It does, on the other hand, have quite a lot to do with your brain chemistry, and with the way highly processed, sugar-rich, fat-intensive, perfectly salty foods make our neurons fire and prompt our brains to release endorphins and other chemicals that make us feel *amazing*.

That's because the taste of highly stimulating foods has an effect on your opioid circuitry, which is the part of your brain that tells you what is pleasurable. Whenever you do something pleasurable – exercising, falling in love, having sex or eating certain foods – your body produces those 'I LIKE THIS' chemicals. Your brain loves endorphins, and you love the way they make you feel, so when you eat foods that trigger this response, even though you are out of control, overeating and probably freaking out on some level, you feel great (at least for a moment).

THIS IS YOUR BRAIN ON FOOD

We all have a food we just can't resist, and when we encounter our greatest temptations, when we take just a tiny little bite, we can't stop until it's all gone. I had a conversation with David Kessler, MD, author of *The End of Overeating*, an amazing book that taught me a lot about how processed foods tempt us to keep eating long after we're full. We talked about how the foods that we can't say 'no' to usually contain the perfect storm of sugar, fat and salt, because our palates are particularly susceptible to that combination of tastes. Although taste is the factor that really gets us to overeat, we are also influenced by other aspects of a food's yumminess, like textures and scents.

SIDELINING TEMPTATION

It's tough not to be tempted when you're surrounded by piles of junk food, or staring down a menu that lists your favourite dishes and all you have to do is tell the waiter what and how many and he will bring them to you in minutes. So what do we do when food availability threatens the careful stockpile of willpower we have worked so hard to develop?

We develop techniques to AVOID temptation.

- **At home:** Get that food out of the house! Stop buying it. Stop storing it. A pantry full of whole grains and ingredients encourages us to cook a healthful dinner. A pantry full of junk encourages us to stuff our faces full of junk.
- **At restaurants:** If you know you're going to a restaurant, come prepared. Decide what you want to eat at home, before reading the menu, because the descriptions and variety of dishes can be a very alluring siren's call. Do you want fish or chicken or beef or something vegetarian? Walk in knowing. Then read the menu just as much as needed to find the item closest to what you've already chosen, and order that. Or read the menu online, if possible, and scan for healthy options. Or if the chef is amenable and they aren't very busy, when the server asks you what you'd like to eat, you can ask if they can grill you up some fish, with some roasted, grilled, sautéed or steamed vegetables.
- **At someone else's home:** Don't show up starving! Have a little snack an hour or so before you arrive at your dinner party so that you don't eat the whole bread basket when you sit down to the table. And do your best to choose healthy options without being swayed by the sights and smells once you're there.

The next time you find yourself eating something uncontrollably, take a closer look at your plate. I bet that the foods you can't get enough of taste sweet and fatty, like biscuits, ice cream, cakes and pastries, even though they are really sweet and fatty AND salty. Or they've been fried and then sauced up with something cheesy or creamy, and they taste salty and fatty, like chips, nachos, and crisps, but they're really salty and fatty AND sweet, because there's some hidden added sugar in the sauce or seasoning. And of course there are the obvious ones that really throw us off the deep end: salted caramel ice cream, chocolate-covered pretzels, maple-cured bacon. That's because the combination of sugar-fat-salt is irresistible to humans, and food manufacturers know it.

Then add into this glorious taste party the other sensory factors, like texture and temperature and scent. The crunch of the biscuit, the smoothness and coldness of the ice cream, the sweet smell of warm cinnamon rolls, the gooiness of the neon-orange cheese dip.

When we encounter these tantalizing foods, a neuronal fireworks display goes off in our brains. We all have neurons that are specifically encoded to respond to single sensory characteristics of food. So, for example, there are neurons that respond to taste and neurons that respond to texture; others respond to the sight, the smell or the temperature of food. And furthermore, there are neurons that are specifically responsive to sweet, salty, sour or bitter tastes. When you eat any food, a chemical reaction takes place that connects each individual message from each neuron, so when you eat highly processed foods, all those different neurons are talking to each other and making the desire to keep eating intense and virtually uncontrollable.

PRACTICE, PRACTICE, PRACTICE

How do you get moving towards health? Practice. You practice making the right choices, again and again, until it becomes second nature. And information helps, because it shows us WHY we are practicing and WHAT our goals should be.

Now that you know that there is a chemical reaction taking place inside your brain *every time* you eat, you can use that information to support the way you practice being healthy.

Think about it: you've had the personal experience of wanting more. You've learnt about the biology behind that desire for more. You understand the difference between being hungry and having your appetite stimulated. So when you realize that the sensations you are experiencing are about appetite and not hunger, you can use your consciousness and awareness of your actions to the highest degree.

AWARENESS: You have to be aware of what is sitting in front of you. And you need to be aware of and totally honest about whether or not you are reaching for it, because it is too easy to sit there in front of a bowl of buttery popcorn and suddenly realize it's all gone. (This has happened to me more than a few times.)

DISCIPLINE: Don't do it. Usually 'just do it' is one of my favourite expressions, but here I say *don't do it*, until you have built up enough willpower to actually stop after one or two bites. How do you build that discipline? I like to come at it from the mentality of what I'm giving myself by NOT eating it.

GOALS: If your goal is to be healthy and fit and give your body the nutrition it needs, and eating the item in question will take you off that course, then consider what matters more to you: eating that food or reaching your goal. If reaching your goal matters more, then by NOT eating it, you are actually GIVING yourself what you want. This kind of thinking makes me feel really good about sticking to my discipline, because I am a girl who likes to get what she wants.

CONSISTENCY: If you usually eat cake five times a week and you've decided to cut down to just once, when you are confronted by a lovely pale yellow slice of Victoria sponge cake or the pretty slice of lemon drizzle cake, do the maths. Have you eaten cake zero times this week? Then once still fits in with your plan. Have you eaten cake every day for the last three days? Then your consistency is out of balance, and saying no is the only way to get that consistency back. The more times that you say no over yes, the more discipline you are building.

PERSONAL RESPONSIBILITY: You and only you can make this choice. You have to decide that *not* eating it is your priority because you want to be healthy, or that tasting something delicious is your priority, because your health is less important to you than immediate gratification. I know that sounds harsh, but it's the truth.

Have patience with yourself and your ability to start changing your food environment from a world of sickly, addictive foods into a nutrient-abundant place where the temptations are healthful rather than life-threatening. It takes time and, most of all, practice.

As you keep practicing, you'll get better and better, and it will become easier and easier.

CHAPTER 26

UNPACKING
YOUR HABITS

W E DON'T CHOOSE OUR instincts. And we don't always choose our habits. But as adults, it is our job to understand *why* we do what we do, whether those 'whys' come from our genes or from the choices we have become accustomed to making, day after day.

As babies, we are all instinct. As adults, we are a motley collection of what we have learnt from the world in addition to our instincts. When you were six months old, your instincts told you to cry when you were hungry. As you got older, you learnt how to ask for what you wanted, discover what you liked and eat with a fork and knife. Today those skills and those food choices have become habits, all of which we are barely aware of as we sit down to every meal, every day.

And habits are a big deal, because they inform everything we do, including the ways we define ourselves. Are you a swimmer? That's probably because somebody taught you how to swim or you signed up for lessons when you were young, and now you grab your suit and head to the pool so regularly that swimming is a part of you. You allowed swimming to shape your schedule and your choices. And in turn, it shaped you, literally transforming your body so that you can glide through the water with power and strength.

But if you had never had the opportunity to dive into a pool, you'd be something else. If you ran daily, you'd be a runner. If you wrote daily, you'd be a writer. If you wanted to be an expert javelin thrower or lion tamer, you'd just have to spend enough time throwing javelins or teaching lions how to jump through hoops.

Same goes for being healthy. Healthy is not an accident, a gift or a rabbit's foot. It is a HABIT. It is a habit that will shape the bodies that we were born with, a habit that will support the genetic makeup we got from our parents. We are not, after all, all the same. Some of us are taller, some are curvier, some have heart conditions or breathing conditions that are not related to eating processed foods. But we are all vulnerable to the consequences of things like bike accidents or exposure to the flu. So if we want to give ourselves a fighting chance to survive and thrive despite our bad backs, our asthma, our eczema, our fallen arches, our dodgy knees and our broken toes, we must develop healthy habits that keep our systems operating at maximum.

WHAT ARE HABITS?

Yeah yeah, Cameron, I hear you saying. *I get what habits are. Brushing your teeth is a good habit. Biting your nails is a bad habit.*

Well, sure, those are examples of habits in action. But do you know how habits actually get created? I didn't, but I asked around and read some great books, like Charles Duhigg's *The Power of Habit*, and here's what I found out: *habits are the way our brains perform familiar activities unconsciously, so we can save energy for other tasks.* What that means is that you can brush your teeth while thinking about what you're going to wear to work. Which is great! Who wants to spend five minutes thinking: Toothbrush. Toothpaste cap off. Toothpaste on brush. Toothpaste cap on. Put brush in mouth. And so on. Unless you're a mime, there's really no need to break it down step-by-step once you've done it a few times.

But consider this scenario: On your way home from work you stop at the local ice cream shop for a cone, which is a little reward you give yourself every day. Maybe when you were a kid, your parents made an ice cream cone your special treat when you did really well in school, so now, you treat yourself. Which is nice, right? We all deserve treats. Except that the first time, it *was* a treat. Then it became a bit more regular. Then it was just a part of your day. Now you might laughingly describe ice cream as 'your favourite' and yourself as an 'ice cream lover'. It's just like running or swimming: When you consciously choose to do something repeatedly that makes you feel good (whether that good feeling comes from adrenaline or sugar or something else), it becomes a habit. In this case, an ice cream habit.

Habits are so natural to our brain that we don't even realize when we've lost control of the conscious decision-making part of the equation. So if you want to change the way you feel, if you want to become healthier, you must unpack your habits. You must wake up to what you are doing now and how it is affecting you. Part of building new habits is making better choices. And part of making better choices is understanding the places where we have stopped choosing, reassessing the passive decisions we've been making habitually and making new, conscious choices that support our long-term goals.

So what I'd like you to do is to take a step back from your habits and recognize that they are CHOICES. Then ask yourself if you like the result of those choices. If not – don't panic, because the good news is, once you become aware of your ingrained choices, you have the POWER TO CHANGE THEM.

But first you have to take responsibility. You have to recognize what your habits are if you want to engage your discipline to change them. You have to wake up.

BEFORE YOU BLAME YOUR BODY, CHECK YOUR HABITS

It's amazing to think that some of the things we believe are totally harmless, or even helpful to us, are the very things that are hurting us. Over and over again, I have seen how some of my daily habits and 'treats' were the cause of symptoms I blamed my body for.

For instance, my dairy habit. I love cheese. String cheese, Cheddar, goat's, Parmesan, Gouda, feta, blue, double-triple-cream Brie. I love that cheese. I always thought it was doing double duty: giving me a dose of protein and calcium along with a delicious snack. And milk – you could often find me drinking straight out of the litre jug. Then I was introduced to the latte, which made milk even more amazing, if that was possible. It's warm, it's steamy, it's cosy, it's caffeinated . . . it's exactly what you want when you're filming a movie, outside, in Boston, in the dead of winter. Believe me, I *loooooooved* my venti latte. It was a newfound dairy treat that I could not put down.

Meanwhile, I was having some stomach issues. My belly was bloated. I was gassy. But this was normal for me: there were often times during the day when I felt bloated and gassy. No matter how many sit-ups I did, I always had a bloated belly, and so I eventually thought, well, this is how I'm built. This is just how my body is. This is just me. But I hated it. I hated feeling bloated. I hated feeling gassy. It made me feel like crap and it ruined my day.

One day, I was online to order a latte with a girlfriend who has a very holistic frame of mind. An hour later, when I started complaining that I felt terrible, she put one and one together for me.

"It's probably from your latte," she said.

"No way," I said. "I always drink these. I'm just bloated, it happens."

"Sure," she said. "It happens when you drink milk."

That didn't make any sense to me. How could something so comforting and seemingly healthy be hurting me? If she was right, what would that mean for me? Would I have to give it up? But my stomach was really hurting. I was desperate. So I put the latte down and agreed to take a break for a month just to see what would happen.

I paid attention to how my body felt over the course of the next week. The bloating was significantly less without the milky drinks. But because I was paying attention, I was able to notice that every time I would have my cheese snack, my stomach would bloat up again. Not as much as with the latte, but enough to be annoying.

That made me rethink my cheese habit, which was a big pain because it was always my go-to snack. So I started experimenting. I completely eliminated cheese and all dairy products from my diet for two weeks. And there it was: flat stomach. No problems. No indigestion. No gas.

Then I thought, *I'm just going to have one latte. I'm going to try it. I'm going to do an experiment.*

You can guess the results. That was when I said, "It isn't worth it. This is *not* worth it." Feeling good meant more to me than eating food that I thought tasted good, food I thought I needed to comfort me or make me happy.

THE HABIT LOOP

Habits have three stages: the Cue, the Routine and the Reward. In Chapter 25 we talked about how seeing an ad for food prompts your appetite. That ad is a cue. Sitting down on the couch to watch TV could also be a cue. Cues are often followed by routines, in which we give in to the urge inspired by the cue. The reward is just what it sounds like – the payoff we get from indulging in our favourite routines.

Here's how it went down in the case of my latte fixation:

CUE: When I felt cold, I began to crave the warmth of a latte.

ROUTINE: How and where I got my latte, getting my hands on it and drinking it.

REWARD: The warmth and comfort I experienced while drinking it, and the feeling of satisfaction that I was taking care of myself.

Over time, that chain of events – feel cold, get latte, feel cosy – became encoded into my brain. Remember how those endorphins get released when your brain encounters sugar, fat and salt? Well, that same neuronal system was being activated every time I completed a habit loop and indulged in my desire to feel cozy and cared for.

In the case of the latte, once I realized it wasn't good for me and I wanted to change that habit, I needed a substitute. That cue – feeling cold, needing warmth and cosiness – wasn't going anywhere, at least until the movie was over. I still wanted the reward – that warm feeling I had when I held and drank from that cup, letting the heat seep through to my gloved hands.

In order to change this habit, I needed to wake up, to take responsibility, use my discipline and develop a new routine. I woke up because a friend woke me up. I took responsibility and identified what was going on when I conducted my no-dairy experiment. And I developed a new routine, and used my discipline to make a different choice whenever I felt cold.

In this case, the new routine was a smaller decaf soya latte. I still satisfied my cue. I still got my reward. And I got the ultimate reward . . . by unpacking my habit and replacing an unhealthy routine with one that worked better for my body, I created a new habit that supported my goals.

SOME THINGS TO REMEMBER ABOUT HABITS

HABITS CAN SNEAK UP ON YOU. My warm, comfy, cosy, belly-bloating daily latte habit crept up on me unawares. One latte, two latte, three latte, four. And boom, I had a latte habit, without even realizing that I was making a choice or what the implications of that choice might be.

HABITS CAN BE SHIFTED. When I realized that I had a habit that was harming me, I was able to consciously create a new habit: the smaller decaf soya latte. Same comfort, no side-effects. Because habits can also be *unchosen*, or shifted, you can create a modified version of your old habit that satisfies the same cravings.

HABITS CAN BE CHOSEN. Like my sunscreen habit. I grew up in the sunshine, but I didn't always wear sunscreen every day. Then I learnt about how quickly the sun can damage my skin, so I started to become more mindful about protecting myself. At first, it was a conscious choice, and I would have to remind myself to do it. So I made sure to always have sunscreen around, the kind that I liked, so it would be easy to apply. Eventually it became routine. Now I am a 'person who wears sunscreen'. It is a habit. A useful, positive habit that cumulatively, done every day, will protect me from sun damage.

The trick to escaping your harmful habits is to identify the ones that have emerged without your say-so and replace them with healthful habits that are intentional and will support your goals. Nowadays, if you're going to the coffee shop and you ask me if I want anything, I don't even have to think. "Sure," I say. "A decaf soya latte; a small, please."

Because I have a new habit. One that doesn't make my body feel bad. And I feel really good about that.

CHAPTER 27

NOTHING'S FOR FREE

R ECENTLY, I WATCHED A group of workmen build a deck over the course of a week. First they installed the foundation. Then the planks. Then the finish. When it was done, all we could see was the shiny wooden surface. But the only reason we didn't fall through that deck once we tested it out was because of the foundation beneath it. The foundation supports the structure. If foundation wasn't laid properly and it wasn't strong, and it wasn't balanced, and it wasn't level, there would have been nothing to stand on.

Discipline is like your foundation for life. It's what supports you and provides stability and structure for everything that you do.

DEFINING DISCIPLINE

I've met a lot of people who think that the world owes them something: a cushy job, a plush office, the right to phone it in when they don't feel like working hard. Those are not people whom I respect. A sense of entitlement is the worst trait I can think of in a person. If you haven't earned it, you don't deserve it.

People talk a lot about the 'secret' to success. But the secret to success isn't a secret at all! At the core of every successful person there is one common thing: discipline. When I think how I define success in my life, I know that strength is important to me and being capable is important to me and feeling good is important to me. I could not achieve any of those things without being disciplined! Everything I have comes from discipline. I do what I do because I make myself do it – *every day* – whether I'm working on a film or on a break between projects. My work is not just my *job*: my work is all the things that I need to do in order to create the life that I want to live.

I can't think of anything that anyone has ever accomplished without having some sort of self-discipline. Without knowing how to work for it. Without learning how to earn it. I talk to my friends who are writers. I say, "Well, how do you do it?" Most all of them will say, "I sit down. I force myself every day to sit down and write for at least two hours. Whether something comes out of it or doesn't come out of it, whether I finish fifty pages or two, I sit down and I do that because I have to make myself do it."

That's what a work ethic is. Any person I know who is successful in my business or any other business is so because they work their arse off for it, because nothing is for free. If you want something, if you want to achieve success in any area of life, you must apply your discipline and your work ethic. Because discipline is what helps you consciously do things in order to reach a desired goal. Discipline is a rejection of entitlement and expectation. Discipline is having a strong awareness that *your choices* have impact and that *your actions* make a difference.

DISCIPLINE IS THE ENGINE THAT GETS YOU THERE

The overall theme of this book is creating whole-body health: mental clarity and discipline, physical well-being and strength, and emotional balance. The way you're going to get there is by harnessing your discipline to put into action all of the knowledge and awareness you've gained so that you can start making smarter choices about what you eat and how you move.

Discipline is not the strict taskmaster keeping you away from your favourite food or forcing you to exercise when you'd rather be napping. Discipline is the force that gives you purpose, focus, strength and determination so that you can accomplish whatever your heart desires. And I mean the productive things that make you feel good about yourself, the healthy things that make you feel energized and alive, the things that *connect you to yourself*. Not the instant gratification you get from inhaling a scoop of ice cream or clocking out before your work is done, but the sense of power and happiness that comes from DEDICATION TO YOURSELF. And part of that is seeing the job or task through to the end.

My father always told me how important it is to do things the right way the first time, so that you don't waste your time having to do them all over again. This was a huge lesson that I still carry with me, because it taught me to be aware of what I was doing, to think about how to do it right and to do it right – because a job isn't done until it's done right, and I sure didn't want to have to do it again.

My mother always told me that nothing is for free. I remember once, as a five-year-old, gleefully pulling a toy out of a box of cereal and saying, "Look, mummy! We got this for free!"

She said, "Honey, it's not free. We had to buy that box of cereal in order to get that toy."

That always stuck with me. Everything you have, everything you get takes work. My parents worked hard to earn money from their jobs so that we could buy that cereal and the toy that came with it. I wasn't entitled to it. And over the years, it was their discipline and work ethic and attitudes against entitlement that showed me the value of hard work and earning what I have. And that translates to every part of my life. Especially when it comes to the most important currency of all, my health.

APPLYING YOUR DISCIPLINE TO YOUR HEALTH

Every day, you wake up and you do things. Some things you have to do: you go to work, go to school, brush your teeth, take out the rubbish, walk the dog. Some things you want to do: read fashion magazines, hang out with your friends or go out to dinner.

If you've been hesitant to make your health a priority, you may want to think about all of the things you apply your energy to, then try to understand what motivates you. Is it a habitual behaviour that you don't even think about? Do you do it because of financial gain? To achieve the respect of your peers? Is it something that you feel so passionate about, so connected to, that you can't imagine NOT doing it?

Whatever your biggest motivators are, try to apply those same principles to the idea of taking care of your health, of giving yourself the right nutrition, of integrating regular movement and training into your life. If responsibility motivates you, remember that it is *your* job to care for your health. If passion motivates you, remember how *amazing* it feels to sweat.

And if financial gain motivates you, consider this: the cost of chronic illnesses caused by bad nutrition and lack of exercise is much steeper than a pair of trainers and healthy groceries.

You already have discipline! I know that because you feed yourself and dress yourself and get yourself from point A to point B. If you didn't have discipline, you wouldn't be reading this book. Discipline isn't a foreign land that you need to discover. It is something that lives within you that you need to *uncover*.

When you accept that the world doesn't owe you anything and that choices have consequences, when you understand that life makes choices *for you* when you hand over the controls . . . that's when life really begins.

If you can harness your energy and attach it to your health, and harness your discipline and attach it to those activities that support your health, then taking care of your health will eventually become just another thing that you are in the habit of doing on a daily basis.

You are capable of that! You know you are. Deep down inside of you, you know that you can do that. And you must, because as much fun as it is to read fashion magazines or go out to dinner, there is nothing else in life that is more important than your health. NOTHING.

You are worth the effort. You are worth the energy.

YOUR PERSONAL STANDARD COUNTS

My discipline began with my parents, who always taught me to do my best. *My* best. Not yours, not anyone else's. Just mine. Same goes for you. You don't have to be *the* best; you just have to do *your* best. You know what that is. And you're responsible for holding yourself accountable to that standard. Discipline requires that you call yourself out. You must be self-accountable! Because discipline does not ask, *Is that good enough for others?* It asks, *Is that good enough for me?*

Discipline and accountability were instilled in me from a young age, and I rely on that foundation every day. Discipline was my mother waking me up in the morning to make breakfast and making sure that I did the dishes and cleaned up after myself when I was done. It was my father teaching me to be accountable for my everyday chores around the house. It was both of my parents instilling in me the knowledge that whatever job I needed to do, I was the one responsible for the result – no one else.

Later on in life, as an adult, I learnt about the value of discipline from other sources, like Master Cheung-yan Yuen. My kung fu teacher from *Charlie's Angels* is the man who helped me discover how to use discipline to discover the real power of my body, and I will forever be grateful to him for that gift.

My discipline is a part of me. It can be a part of you, too. Whether you learnt discipline as a child or a young woman, or you are just discovering its power now, cultivate your pride and self-accountability. Get to know yourself well enough to know what you need to be successful. Create a long-term plan that uses short-term actions to reach your goals. Follow through and be consistent. All of these actions are how you cultivate discipline: identifying what needs to be done, setting your intention, taking conscious action, following through and doing your best.

When you accept that the world doesn't owe you anything and that choices have consequences, when you understand that life makes choices *for you* when you hand over the controls and that you must work hard to get that power back, when you have identified what the right choices are and developed the discipline that will enable you to consistently make those choices. That's when life really begins. Discipline isn't about denying yourself; it's about giving to yourself. It isn't about losing; it's about gaining. Every day that I get up early in the morning to go to the gym, even if it's difficult and I'd rather stay in bed, I think about what will I gain from going to the gym and what will I lose if I don't. Nine times out of ten, I will only gain by going and lose by not going. Discipline is always a gain in the tallying of life.

Discipline builds discipline. It's like a muscle: the more you use it, the stronger it becomes and the further it can take you.

CHAPTER 28

PLANNING FOR YOUR NUTRITION

———

EATING WELL IS YOUR responsibility, no matter who you are or where you are. So is incorporating movement into your life.

When I'm working on a film, we're on set from dawn till night. At the same time that I have to be conscientious about my movement – running everywhere I can, just to keep my blood moving and my energy up – I also have to be on top of my nutrition.

If you know about how movies work, that might surprise you, because contractually, everybody on set needs to be fed – so why would I need to think about what I was going to eat if food is available? Well the kind folks in craft services – the outfit that feeds everyone on a set – lay out a full spread that the crew can graze throughout the day, in between the scheduled meals, which come every six hours. And this spread is intense. Picture a huge buffet table covered with every kind of food imaginable. Bagels and muffins and doughnuts. Biscuits. Cheese, cream cheese, yoghurt, fruit, more biscuits. A mile full of the kind of food that is easy to lay out on a buffet table. The kind of food that is easy to snack on. The kind of food that makes people complain about how much food they've eaten and how gross they feel. 'Body by craft' is what we call it on set.

Most people don't even think about it. They don't take responsibility. They just eat the food, feel terrible, gain weight and blame the food. It's an easy trap to fall into. The food has been provided so that you don't have to think about it, because when you're on a set in the middle of nowhere for twelve hours a day, you're gonna have to eat something. And although there are usually some healthy options on the table, it's hard to restrict your appetite when there's such a huge variety of so many things, right there in front of you. What used to be set up as a courtesy to the crew has now become a potential time bomb for many of the crew members' health. When you work as hard as they do, you get hungry, and there is little time to step away and really be thoughtful about what you're eating. Situations like these create an atmosphere that becomes the perfect storm of bad habits, poor nutrition and mindless eating.

Again, it comes down to personal responsibility. You can blame the environment of tempting food that surrounds you, or you can accept that you have the choice of what to eat and how much to eat. To be conscious and aware of the food you put in your body. You can't use being away from home as an excuse: when you're out and about, it is your job to make the best possible choices. Whether you're sequestered for jury duty or at an office or in a friend's house or on a movie set, you can make better choices. Remember: hunger is your friend. So plan for your hunger so that you can feed and nourish it instead of temporarily squelching it. You must take ownership of it, take responsibility for it and then you must do it.

START WITH BREAKFAST

I've had the same makeup artist for the past ten years. Her name is Robin. As you can imagine, after a decade of having an up-close and personal look at this face of mine, she knows my skin really well. Often, Robin and I have to get started early in the morning – and I mean early. A 5:30 am call time on set means that I've got to wake up at 4:00 am to take a shower and get to the set by 4:30.

If I haven't found time to eat breakfast before I sit down in front of Robin, she can tell right away. She'll take one look at me and say, "You haven't eaten yet."

She just knows!

I'll say, "I know, I know. The food is on the way."

And she'll say, "OK. Well, let's just wait until it gets here and you take a few bites."

She will refuse to do my makeup until I eat, because as soon as I take two bites, three bites, my skin changes. Then my skin can hold the makeup. Then Robin will look at me under the lights and declare that we are ready to start.

She knows that breakfast is the most important meal of the day because while we sleep, our bodies are in a state of rest, repair and recharge. When we rise and want to shine, we must give our bodies nutrients and energy again so that it can do all of the things we ask of it. So we BREAK our night of FASTING by eating BREAKFAST.

Eating breakfast is essential because it helps you . . .

- Reach your daily nutritional requirements.
- Maintain a healthy weight (people who skip breakfast tend to overeat at other meals).
- Build healthy habits that lay the foundation for other good nutritional habits, because planning for a solid breakfast every day requires discipline.

I always plan for my breakfast, because the quicker it is for me to prepare my morning meal, the easier it is for me to keep this daily habit. That's what I'd like for you: to plan your breakfast so you get the nutrition you need, first thing. Not a sugary breakfast you grab on the go, but a real, wholesome breakfast you prepare yourself whenever possible. Mornings tend to be the busiest time of the day for a lot of people, but that doesn't make eating nutritious food any less important. In fact it makes it even more important.

SETTING YOURSELF UP FOR SUCCESS

We all know ourselves well enough to have an idea of when we will get hungry during the day. You know if your stomach typically starts to rumble in the late afternoon. You know that a change in your eating habits – like going to an eight pm dinner when you usually eat at six – means that you'll need a snack to get you through. You know that if you have to get up extra early to drop off the kids at an activity, you'll be temped to hit the drive-thru by the time you're on the way home.

So how do you set yourself up for success when it comes to your health? The same way you plan for success when you're packing for a holiday: if you know you're going to go swimming, you bring a bathing suit. If you know you're going to go hiking, you pack your hiking boots. It's called forethought, planning, preparing. Same goes for your nutrition. If you pack a healthy snack or breakfast, or even identify places where you can pick up a healthy snack or breakfast, you're avoiding making a bad decision once you're hungry, and you're giving yourself the fuel you need to actually feel good and stay focussed. Setting yourself up for success keeps you conscious of the choices in the moment, so that you can be ready for those inevitable moments when all of your willpower must be summoned to keep you moving forwards to your goal.

Planning ahead is the most useful tool I have when it comes to my nutrition. It assures me that I will be prepared no matter where I go, with food that meets my standards. It ensures that I'm never stuck somewhere with no other options except ones that are *less* bad instead of options that are *more* nutritious.

SHOPPING AND COOKING FOR THE WEEK

In my house, on Sundays, you can usually find me in the kitchen getting my food ready for the week. I tell my friends that I can't meet for lunch; I tell my nieces and nephews that if they want to hang out, it will have to be at my house. I make sure not to have any meetings or calls scheduled during that time, because that time is SO IMPORTANT for me. It's the time that sets me up for a successful week, for feeling good, for knowing that I am taking care of myself.

If you can get a few hours clumped together, you can do your shopping and your cooking in one swoop. If you don't, you may have to plan a grocery trip on Saturday and your cooking on Sunday, or go shopping on Sunday and cook on Monday evening . . . whatever works for you.

My vegetarian friends might make one vegetable soup, one big kale salad that can stay in the fridge without getting mushy (to avoid sogginess, don't dress your salad until you're ready to eat it), and one pot of lentils and rice. That way, they always have a ready snack or meal whenever hunger comes to visit. Sometimes I prepare all the fixings for a salad, store them separately for freshness, and then assemble the salad the night before or morning of. For my protein, I will usually make chicken, which does marvellously in the fridge over a few days, so I will make enough on a Sunday that I have it to hand to slice into my salad for lunch or heat up for breakfast with an egg and some greens and my porridge during the week. There are so many wonderful cookbooks and websites out there that offer fresh, healthy recipe choices. Do some exploring and find a source you like, then make that book or blog your new best friend.

Just remember that everybody plans for success in her own way, but the key is PLANNING. I have friends who never cook with recipes; they're the type to just go to the store and throw heaps of vegetables into their carts, then go home and decide right then and there in front of the chopping board what they want to make. Some friends prefer to operate in a more orderly way, so they'll choose a few recipes, make a list of needed items, follow the shopping list to fill their trolley, and follow the recipes once they get all of the ingredients to the kitchen.

When it comes to your shopping list, think about what you want to accomplish and create a list that matches your goals, your cooking style and what you'll be getting up to over the week.

PLANNING AND SHOPPING

When you sit down to plan your meals and shopping list for the week, here are a few things to consider:

HOW ACTIVE WILL YOU BE THIS WEEK? When I'm planning my meals for the week, I always take into consideration how much physical activity I will be doing. If I know that I'm going to be able to go to the gym five days that week and my training will be at a higher level, then I will plan my meals to support that schedule. But if I know I'm only going to go to the gym and train hard on two days and the other days I'm going to fit in a few light workouts just to get my heart rate up when I can, then I will design my nutrition around my body's needs for that level of activity. Planning ahead for your nutritional needs goes hand in hand with your physical agenda. It's an equation, always.

ARE YOU GOING TO BE PREPARING BREAKFASTS? If you've decided to make breakfast at home in the morning before you leave the house, ask yourself what you might need for the week so that you will have it to hand and can prep it the night before. Eggs? Bananas? Oats? Make sure your kitchen is well stocked.

HOW ABOUT GRAINS? Will you be eating quinoa or brown rice, maybe a little millet? Preparing grains in advance is a lifesaver when you're having a busy week, and they keep well in the fridge. I like to make a couple of batches of different grains for variety. I'll do a batch of brown rice with black lentils, and then a batch of quinoa, and alternate them throughout the week.

Don't forget protein. Will there be meat on the menu? Chicken? Fish? Beef? Or will you eat less meat this week and concentrate on getting your proteins from a variety of beans and rice? Keep some eggs to hand for protein in a pinch, and remember that your pot of rice and lentils fills your protein quota as well as your whole-grains quota.

PLAN YOUR PRODUCE. What vegetables do you want to have to hand for salads, roasting, steaming, sautéing? What kind of fruit do you love to snack on? Get a variety of fruit and vegetables, but be strategic, because fresh produce doesn't last as long as grains and proteins. If you buy kale or tomatoes or spinach for a big salad, you might want to use them early in the week. You can

FRESH, FROZEN OR CANNED?

Fresh veg are the tops. In the summers, fresh vegetables are everywhere – at farmer's markets in the country, at market stalls in the city – but depending on where you live, fresh may not always be accessible. The fresher the better, but if you get to the market and the vegetables are looking a little wilted, don't sweat it. Frozen vegetables are the next best thing.

Some frozen vegetables actually contain more nutrients than fresh vegetables, because they're picked at their peak and flash-frozen, which preserves their nutrient value. Often fresh fruits and vegetables are picked before they're ripe because they have to travel so far to get to you. Not only does this mean less nutrient value in the plant, but the travelling process also exposes them to heat and light that degrade them more quickly. So don't shy away from frozen; just make sure that when you prep them, you steam or sauté instead of boiling, because boiling can drain them of the water-soluble nutrients that they have been retaining for you while frozen.

Tinned vegetables should be your last choice, because tinned food loses a great deal of its nutrient value during processing, and often salt, sugar or other preservatives are added. Try to avoid tinned foods if you can.

plan to eat your heartier veggies, like brussels sprouts, peppers or carrots, later in the week – they'll hold up better in the fridge.

DON'T FORGET THE LEMONS! Lemon juice is a great chef's secret. There are very few things that I've eaten that didn't taste better with a little squeeze of lemon, from meats to vegetables to grains. I like bright flavours, and lemon always makes everything a little brighter. Plus, the acidity from the lemon lends food a slightly salty flavour.

PREPPING YOUR MEALS

For me, my prep time is a very relaxing time that I take for myself. It allows me to be thoughtful about my health and think about what I enjoy eating and nourishing my body with. Sometimes I prep my foods for the week with friends as a little gathering at the weekend; we all cook together and make one dish that we can parcel into individual portions. Or sometimes we might each agree to make extra helpings of something healthy and delicious and then divide up the meals and share all around.

When I'm prepping, I put my batches of two different grains on the hob first, because they can usually be left to cook on their own for at least twenty to forty-five minutes. While they're going, I start to clean my vegetables and cut them into whatever sizes I want for them to be grilled, sautéed, steamed or roasted. Then I'll cook the meat next. Usually I'll make some chicken breasts, nothing fancy, just some garlic and salt and olive oil. I'll grill it or pan-roast it, or roast it in the oven. My habit is to try to make the dishes as simple as I can, because the simplest foods can be the most tasty.

And when all of the prep is done and the food is cooked, I'll make up a few containers, each with a portion of protein, grain and vegetable. I'll pack one for each day and a little extra for that emergency evening when I find out I'm going to be working later than I thought. I feel so good when everything is prepared and ready to be packed into my cold bag for the day or transported to my refrigerator at work. I know that any meeting I go to will be fuelled by the grains and chicken and greens that I pack with me. I know that if I have to get on a plane, I will be able to tell the cabin crew that I don't need that frozen airplane meal because I have my own delicious food with me. I know that I am planning ahead to take care of myself, and that all of the choices I make during the week will be a little easier because of that.

PLANNING TO EAT LESS SUGAR

Identifying all of the added sugar in meals and snacks, particularly if you are interested in losing weight, can be very challenging – but also very rewarding. Go back and read Chapter 7 again carefully, then take a walk through your kitchen and pantry and start reading the labels. How many of the foods that you consume on a daily basis have added sugars you didn't even suspect were in there? What simple swaps can you make to reduce your intake of added sugar? Here are ten easy ideas:

1. Replace sweetened bottled teas and instant tea powders with unsweetened iced tea and tea bags
2. Learn to enjoy your coffee or tea without adding sugar.
3. Buy natural peanut butter instead of sugar-added peanut butters.
4. Swap your sugary cereal for a cereal with no sugar added, or plain oatmeal.
5. Go for plain yoghurt and add your own fresh fruit instead of buying fruit-on-the-bottom yoghurt.
6. Make your own salad dressing by whisking together lemon juice, or any vinegar you like, with a little olive oil, instead of buying the bottled stuff.
7. Lose the sweets. Eat the berries.
8. Choose mustard, vinegar or pesto instead of ketchup.
9. Swap canned tomatoes for spaghetti sauce.
10. No more squash!! Add lemon, lime and mint to sparkling water instead.

After you've gone a week without added sugars, try a sip of sweetened iced tea. Does it taste sweeter than you remember? Giving your taste buds a break allows you to really taste all of the flavours in food in a new way, and you may be shocked by just how sweet the foods are that you've been eating!

REMEMBER TO EAT WHEN YOU ARE HUNGRY!

Eating when you are hungry is one of your main jobs as a human. It's funny how we seem to forget that. If you read magazines or have best friends who don't really know a lot about nutrition even though they always seem to be 'on a diet', it's easy to think that hunger is something to battle rather than embrace. As we've already discussed, it's essential to *feed your hunger*. Remember, hunger is not appetite. When you feel your body asking you for fuel, consider the following:

- **Know what hunger really is.** Hunger is a sign that you need to eat real, nutrient-rich food so that your body can keep on keeping on. If you are accustomed to overeating, it may take some time getting used to eating smaller portions before your internal signals start to make sense to you. Trust yourself. Trust your body. If you ate a healthy meal, you should start to feel hungry about three hours later. Do NOT wait until you are a raving maniac to eat. Eat when you feel that first tap on the shoulder – and eat lightly, until the tapping stops and you feel normal again. Not full. Not hungry. Just . . . satisfied.
- **Recognize fullness.** The term *satiety* is used to describe the physiological and psychological experience of fullness that comes after eating and/or drinking. When your body is full, what it is saying to you is that you've given it enough fuel to keep calm and carry on. Once you recognize that feeling – which is an absence of hunger more than a feeling that you need to unbutton your trousers – you can stop eating.
- **Focus on 'mindful' versus 'mindless' eating.** When you eat mindfully, you eat without distractions (no TV, smartphone, computer, magazine or newspaper). You eat slowly and pay attention to every mouthful. You don't accidentally consume a bag of pretzels while watching a particularly engrossing episode of this week's favourite TV show.

Here are three steps to practice eating mindfully:

1. Before your meal: Instead of grazing mindlessly on whatever food is available to you, wait to satisfy your hunger and choose foods that you like and that will give you the fuel you need.

2. During your meal: Eat slowly and focus on the enjoyment of your meal. How does the food smell? How does it taste? Is it crunchy? Smooth? Spicy?

3. After your meal: Consider how the food you ate is making you feel. Do you feel alert? Sluggish? Energized? Bloated? Is it a feeling you want to have again? Is it a food you want to eat again?

Being mindful about your eating begins with planning and prep, continues through packing up meals and taking them with you, and extends all the way until the last bite that leaves you satisfied, not stuffed. Because all of that planning is there for one purpose: so that you can feed yourself well and easily whenever you get hungry, and have the energy you need to fuel your life and your movement.

CHAPTER 29

FINDING YOUR INNER ATHLETE

YOU ARE AN ATHLETE. Yes, you. Even if you've never run a mile, let alone a marathon. Even if you've never done a push-up, let alone twenty. Even if you've never thrown a ball, hit a ball for six, or kicked a ball into a net, you are an athlete who just hasn't reached her full potential. Yet.

You know what makes people feel athletic? Athletic activities.

Being an athlete is not about winning a contest or getting a trophy. It's about discovering the joy of moving your body and pushing its limits, the triumphs of discovering those first knotty muscles in an arm that used to feel as weak as custard. That triumphant sensation that comes not from winning but from WORKING.

I have a dear friend, a little pipsqueak, who had never done anything athletic in the fifteen years I'd known her. One day she decided to try a group fitness class. So she went to the class, and she liked the way she felt afterwards, so she signed up for another one. And then another one. As she kept training, she gained strength, and she realized that she couldn't be without her workout: she just LOVED that feeling of moving, and loved that feeling of becoming stronger. Once she gave herself the chance to find her inner athlete, once she realized that life is better when you are active, she made a conscious choice to prioritize fitness. Now, three years later, it's become a habit. She works out four times a week, really challenging, athletic workouts, and she has transformed her body and her mind – she has more emotional and mental strength than ever before. And this is a busy lady. She runs a business, she has two children. She's always been disciplined and had a work ethic – and now she applies that discipline and work ethic to her own health and fitness, and she is positively thriving.

She made a choice, and she gave herself a chance, and she changed the game. Don't you want that same chance?

GET MOVING

Here's how I begin every single day: I wake up, drink water, fuel up, and MOVE. As soon as I brush my teeth, I begin to wake up my system. I drink my litre of water to get things going. I have a quick bite – some leftovers from the night before or some quinoa and lentils, just to give me a bit of fuel so that I can do what I need to do before I prepare my big breakfast.

That's what I want you to do: wake up, fuel up, get moving. If you don't have time for a full workout, you'll plan your workout for later, and give your body a good wake-up call by jumping up and down until your face feels warm, doing some squats, a sixty-second plank. Or hit the pavement and run around the block as many times as it takes to get the blood flowing. That's it! Sweating is ideal in the morning, even if it means you have to wake up early.

By starting your day with movement, by making a strong choice daily, first thing in the morning, you're also strengthening your discipline. It shows you that you can do it. That you should do it. That you must do it!!

PLANNING FOR COMMITMENT

Just as you plan for your nutrition, you need to plan for your movement.

The excuse that I hear most often about avoiding exercise is, "I don't have enough time." I get that we're all busy, but sometimes we don't make the best use of our time. Anytime you are trying to be prepared, the mentality should always be, "How can I save time *to help me reach my goal*?" Sometimes setting yourself up for success isn't glamorous. It means doing things like laying out your clothing, packing your lunch and snacks for the next day and getting to bed at a decent hour.

Planning helps you use your discipline and create new habits, and it ensures that you'll have the time to work out and the equipment you'll need to do so. Thinking ahead saves time when you are trying to fit everything into your schedule. Here are a few ways I help myself keep on track with my fitness:

WRITE IT DOWN. When you have a doctor's appointment, a meeting or a date, you always manage to show up, don't you? That's because you schedule it, you take it seriously and you write it down. Your fitness deserves the same respect. So plan your fitness, schedule it and write it down.

LAY OUT YOUR CLOTHES. For morning workouts, lay out your gym clothes and your after-gym clothes the night before. If you're not going home after your workout, pack the clothes and supplies you'll need to shower at the gym. Personally, I find it less frustrating to make decisions the day or night before rather than at five thirty in the morning. It's inevitable that I'll forget something if I try to make decisions that early.

MAKE A TOILETRY BAG FOR THE GYM. I keep a small travel-size toiletry bag in my gym bag that has all the items that I need to take a shower and prepare for the day ahead. That way I don't have to remember every single thing I need every time I pack my bag. It's already there.

CARRY EVERYTHING AROUND IF YOU NEED TO. Sure, it can be annoying to schlep an extra bag. But it's less annoying than having to forgo your workout because you can't fit in the time to run home and change. If I have a workout scheduled for the end of the day, I'll pack a bag and keep it with me until it's time to get my sweat on.

USE ACCESSORIES TO YOUR ADVANTAGE. If I know I have to go out at night and I'll be going to the gym beforehand, I'll find a way to turn my day outfit into something evening appropriate by bringing a change of shoes, a small bag, or jewellery that works for the evening. That way my gym bag doesn't overwhelm me. On days like this, I also try to make my gym bag a little more like a large handbag rather than a 'gym bag', so I'm the only one who knows I brought my sweaty gym clothes to a fancy restaurant.

BRING EXTRA UNDERWEAR. I always keep a couple of pairs of clean knickers and a bra in my gym bag. They're the easiest thing to forget to pack! You'll need them when your underwear from a workout is sweaty and you want to change into your street clothes. You will thank yourself for this particular act of planning at some point in the future, I promise you that.

WHAT LANGUAGES WILL YOUR BODY LEARN?

Part of what's so great about moving your body and exploring what it's capable of is that there are endless possibilities. It's like learning a new language – and once your body can speak fluently, you're going to want it to talk all of the time.

Planning helps you use your discipline and create new habits, and it ensures that you'll have the time to work out and the equipment you'll need to do so.

Ever since I discovered that my body could speak 'kung fu' on the set of *Charlie's Angels*, I have found new languages for it to learn. During the first *Charlie's*, I learnt to snowboard, and after the second *Charlie's*, I learnt to surf. I reconnected with my childhood passions for running and hiking. And I found that the more languages I taught my body to speak, the easier it was for it to keep learning new ones. This fluency has given me the opportunity to have a diversity of experiences. I can participate in so many different things because I have the strength and the basis of knowledge in my body.

It's important to match your athletic activities with your interests and your lifestyle if you're going to stay consistent and stick with your programme. When you think about what kind of fitness programme might be right for you, ask yourself these questions:

WHAT ENVIRONMENT INSPIRES YOU? Personally, I love the gym. I love being around people who are sweaty and pushing themselves, and who are all focused on the same goals. I love being in a group like that. When you spend a lot of time working out in an environment with like-minded people, you're bound to meet others who also want to kick their collective arses.

WHAT FITNESS ACTIVITIES ARE YOUR FRIENDS INTO? Do you have friends who love to play tennis? What a fun thing to do on a Saturday afternoon – meet your friends for fresh air and a game of tennis. Or tag along to a yoga class or a spinning class with your neighbour. Fitness buddies are the best. But pick your buddy wisely; try to work out with someone who's roughly at your fitness level, someone who will push you, someone who is reliable and won't flake out on you. And always have a plan B for those times when your partner does cancel. Go for a run if your tennis partner had a last-minute conflict. Their excuse can't be your excuse.

WHAT'S NEARBY? What activities are readily available in your community? Do you live near a lake? Do you live near a great pool? A place to go hiking? A park with a fitness route and chin-up bars? Matching your fitness to local availability helps you stay committed and lets you get to know your community a little better.

HOW MUCH DO YOU WANT TO SPEND? Some people will do anything to sell you their quick-fix miracle cure, as long as you're happy to swipe your credit card. But being fit doesn't need to cost anything, as long as you show up. Anybody can do a push-up. You can do one right now! There's no fee involved. And while bicycles and weights and StairMasters can be useful, you don't *need* them. You don't need anything: just your own body and the desire to make it happen.

GETTING TO FIT

Being fit is fun; getting there isn't easy, but it can be as awesome and inspiring as it is challenging. When you're thinking about starting a training programme, choose an activity that you will ENJOY. Don't choose something because it promises you dramatic results, or because it's trendy, or because it looks cool. If you don't like it, you're not going to stick with it – so none of those things matter. Choose something that feels playful or sporty or outdoorsy. Or something that feels social, upbeat, connected or community oriented. Or something that feels personal, soulful, quiet, meditative and recharges you mentally. However you like to be in life, you should choose a physical activity that reflects those same preferences.

When you were a kid, maybe you played on a local football team. Well, why did you stop playing? If remembering running around on the field makes you happy, consider that there are plenty of football teams around for adults. Part of getting into working out is to stop thinking of it as a chore and start thinking of it as your playtime, when you can let your body move, get your energy levels up, get your heart racing, get excited, all of the things you used to take for granted when you were young. So think about the things that used to bring you joy, and incorporate them into your life now.

Or try something new.

Trying something new can be intimidating. None of us wants to feel defeated on our first jog or mess up in front of a whole yoga class. But who cares? At least you're out there and you're doing it. I've always thought that Nike slogan is genius: *Just do it*. Its genius is in its simplicity. It applies to everything that your brain tries to talk you out of. I say it to myself every day at least five times a day. It really is the answer to anything that you want to do: you have to *just do it*.

Like your job. How did you learn to do your job? If your job is to write reports, decorate cakes or sell property, you had to learn it somewhere. You weren't born knowing about law or Royal icing or estate agencies. The only way you've learnt any skill in life is by *just doing it*. You may not have won first place every time, or even made it to the finals. But that's not the point. If you can do it, it's because you kept doing it.

I mean, think about it: In order to make it to adulthood, we all had to learn things. Maybe you had to study to get into uni, so you went over your lessons and notes again and again until you could pass your tests and continue to learn. Or maybe you learnt how to be street smart on your feet, how to simply survive walking home from school every day without getting caught up by someone who wanted to jack your world. Teaching your body to move works the same way: you just have to do it over and over and over again until it becomes second nature. You had to learn the things that got you to where you are now, and now you have to learn the things that will get you to where you want to go next.

And as soon as you start to use your body, you'll see how it responds. How it leaps to create new strength and understand new ways of moving and being.

MORE GOOD REASONS TO MOVE

No matter how good you think you look in a pair of jeans, you STILL NEED TO TRAIN. Being a size two doesn't mean that your body composition is lean or that you have the kind of muscle and bone strength that will support you as you age. And if you're someone who's already put in the effort to go from heavier to lighter (bravo!), you know that ya gotta keep on sweating to keep it up or else it goes away. No matter how young you are, how thin you are or how much your boyfriend says that he likes you with a little more shimmy to your shake, you must move. This isn't just about how you look in a dress or the look on your man's face when he sees how you fill out that D cup (or A cup). It's about your physical health. Your strength. Your endurance. And a million other things you might not have considered.

AMBIEN AFICIONADOS: When you shake what your mama gave you on a regular basis, you'll find that you sleep more deeply and restfully.

SMARTY-PANTS: When you are active, you are giving your body what it needs to do its myriad important jobs, like circulate oxygen through your blood into all parts of your body, especially to that big beautiful brain of yours, so you can think more clearly and be more productive.

SLEEPYHEADS AND COFFEE JUNKIES: Some people worry that working out makes you tired, and if you're not getting the right nutrition to support your movement, that can be the case. But if you give your body the right food and fluids that it needs to sustain your movement, working out is like a rocket boost for your energy levels.

PREVENTIVE-HEALTH PEOPLE: Do you have family members who suffer from chronic diseases, such as heart disease and type 2 diabetes? Regular exercise decreases your risk for these and other ailments.

GLOOMY GUTS: Working out makes you feel euphoric during and after, with overall mood improvements and a reduction in the incidence or severity of depression and anxiety. That's right! Moving makes you strong, and it makes you HAPPY!

THE MOST IMPORTANT FIFTEEN MINUTES OF YOUR DAY

I love to work out first thing in the morning: it starts me out on the right path and gives me a boost for the rest of my day. Even if I only have twenty minutes, I run up and down my block. Or I put on my headphones and trainers and just dance around the living room. I find that by eating a healthy breakfast and getting my sweat on in the morning, I launch myself into the day with energy instead of yawning my way through meetings. Indoors or outside, even fifteen minutes of movement gives me a jolt of energy and reminds me how good it feels to move.

No matter how good you think you look in a pair of jeans, you STILL NEED TO TRAIN. Being a size two doesn't mean your body composition is lean. . . . Moving is not about how we *look*. It's about how we *feel*.

Because moving is not about how we *look*. It's about how we *feel*. It amazes me that people forget that! If you spend some time eavesdropping in restaurants, in dress shops and at parties, you'll hear women complaining constantly about how they look and how they should probably consider exercising more to fix that. We're always talking about how we look, when that's not the real reason for moving. Moving is imperative for ALL PEOPLE who want to be strong and healthy, who want to live their years out in bodies that can support life instead of getting sick and getting weak.

So please consider starting your day with movement. Just fifteen minutes, if that's all you've got. Use your discipline to give yourself those moments to start the day off right. According to the guys who literally wrote the book on willpower, Roy Baumeister and John Tierney (the authors of *Willpower*), willpower is stronger in the mornings for most people. So why not plan to move at the time when you're most likely to follow through?

Here's a great way to start moving in the morning: Create a fifteen-minute playlist that will make you wanna dance. Put your headphones on. Jump up and down, swing your arms, touch your toes, run in circles, however it makes your body want to move, but don't stop until that playlist is over! Do that every day till you get mad that the fifteen minutes is over. And when that time comes (and it will come), make that playlist into thirty minutes, then into forty-five, and then an hour. And with every additional fifteen minutes you add, you and I will be high-fiving! Because every increment of success is a high-five moment. So high five to you, beautiful lady! Nice work! Keep on going!!

NOW YOU REALLY GOT THIS

N THIS BOOK, THERE is no goal to reach in 7 days or 30 days or 365 days. The goal here is forever. It is not a quick fix. It is about longevity. And what you will earn is measured not in stones or centimetres lost, but in what you will gain: a sharper, clearer mind; a body that can power the actions your mind dreams up for you; a confidence that comes from knowing yourself, caring for yourself, and respecting yourself. The rewards are continuous and evolving; the work is daily. It's about using your discipline to make consistent choices that support your progress towards a goal that *you* have chosen. And the ultimate goal, for us all, is a long, strong, happy, healthy life.

But you can't just read this book and nod along in agreement. Thinking about it isn't the same as taking real action. You must actually commit to it. You must want it. And then you must go after it. Every time you make a better choice, you are strengthening your discipline and changing your habits. These little shifts in awareness will set you up to create more and more healthy habits that will sustain you over a lifetime.

Good health starts with awareness and relies on personal responsibility. It is translating intention into action. I hope what you've learnt about nutrition and fitness has deepened your knowledge and understanding of how your body works and what it needs to survive and thrive.

Even with all of the health knowledge in the world, habits are tough to shift, especially when there's a box of jelly doughnuts on your co-worker's desk. (That is, if jelly doughnuts are your thing. For me it's still chips.)

So here's something for your mind to chew on: just as it has taken time to create the habits you have now, it will take time to change them. And the only way that you find success in doing that is by being kind to yourself. Part of taking responsibility for your health is being kind, loving and supportive to yourself. Part of becoming disciplined is encouraging yourself to get up and keep going. There's a balance to achieve: you need to keep yourself accountable, without beating yourself up if things aren't exactly right all of the time.

Take steps to connect your body and your mind so that you can develop an awareness of how your actions make you feel, and consistently strive towards those actions that truly make you feel good inside and out. Make those actions a habit. Build and reinforce your own work ethic for your own health each day.

A human being eats five times a day. A human being should sweat at least once a day. That gives you six times a day, every single day of your life, to choose to be conscious or unconscious. To choose to be awake or to live in dreamland.

So wake up. Love yourself. Care for yourself.

Your body is the most precious thing that you have.

ACKNOWLEDGEMENTS

THIS IS WHERE I get to thank everybody who made this book possible.

First, my parents . . .

There are way too many things to thank my Mom and Dad for, SO many amazing lessons and tools that they have given me throughout my life, SO many examples of what it is to be good people with big, loving hearts, and devotion to the ones they love. I am the result of their love, dedication, diligence, care, trust, nurturing, patience, guidance and wisdom.

And I'd like to thank my Mommy for all the nurturing hours that we've spent together in the kitchen. It has been one of the greatest gifts that you have given me. You taught me how to infuse love for family and friends into the food that you feed them. I cherish every moment that we have ever spent in the kitchen together. Every meal we have shared together with our family and friends, even every conversation that we have ever had discussing our love of food. For goodness' sake, we spend just about as much time talking about food as eating it. All of those meals cooked, shared or fantasized about not only nourish my body but also nurture my heart and soul.

I love you, Mommy! You're the best in the Biz!

THESE ARE THE PEEPS WHO LITERALLY MADE THIS BOOK HAPPEN . . .

Jesse Lutz for being the boss of me, for working your magic on a second-to-second basis to keep the machine in motion. And for all of your love, your nurturing, your humour, your guidance, your keen observation and your big brilliant brain, which all served this book immensely. And of course for being the best Wifey a gal could ever have.

Ricki Yorn, you're my rock. Thank you for always believing in me. With your guidance and vision you always help me to navigate the right path. I feel so blessed to have you as my partner, my brother and to be on the receiving end of that big, beautiful heart of yours. Thank you for helping me see this book all the way through.

Sandra Bark, what can I say? Without you this book would not exist as it does. You understood what I wanted this book to be, and I thank you for doing

259

all the heavy lifting. Your wisdom, curiosity, way with words and understanding of others, as well as your ability to be honest and candid about yourself, was the perfect formula to fuel your talent and to bring the vision for this book to fruition. I learnt SO much from you. I couldn't have asked for a better partner or teacher – or personal chef!! ;)

Julie Will, for believing in this book, and knowing exactly what needed to be cut and what need to be included. Your knowledge and experience were invaluable in shaping this book, and your spirit is incorporated into its content. Thank you for your partnership.

Brad Cafarelli, NO ONE does it better than YOU, B-Caf!! You see it all, you understand it all, you navigate with grace and integrity. Thank you for your guidance in how to bring this book into the world. As always, you understand the big picture, and you're always on point, and always doing it from the heart, with love and thoughtfulness.

Marcy Morris, for always being fair. For your strength and your guidance. You always make decisions for the right reasons. Your love and dedication have been a blessing in my life.

Nick Styne, even though you weren't directly a part of this book, your love, support and joyful, loving spirit were felt in support of this endeavour. You are the key to it all.

Jennifer Rudolph Walsh, your expertise and guidance brought the right people to this project. Thank you for making it all happen.

THERE WERE SO MANY AMAZING PEOPLE WHO DEDICATED THEMSELVES TO THIS PROJECT, SHARING THEIR TIME AND THEIR ENERGY TO MAKE IT ALL HAPPEN . . .

Thank you so much to all the experts who answered my questions and shared their knowledge. Dr Kathleen Woolf gave us a thorough primer on human nutrition and went over our material repeatedly. Dr Aurelia Nattiv shared her fitness expertise and read over our shoulders to make sure we were as accurate as possible. Dr Diana Chavkin taught us a lot about our lady parts and made sure our information on female health was useful and correct. Dr David Kessler's work on the American appetite inspired me, and so did the conversation we had about this project. Dr Brian Wansink was gracious enough to talk to us about his studies on eating behaviour, and Dr Martin Blaser and Dr Maria Gloria Dominguez-Bello invited us into their offices to share their research about the human microbiome.

I also want to acknowledge the amazing professionals who shared their expertise and dedicated themselves to this project: Karen Rinaldi, Senior VP of HarperWave; Paul Kepple and the team at Headcase Design; illustrator Patrick Morgan; our publicist at Harper, Leslie Cohen; design manager Leah Carlson-Stanisic; and Kathy Schneider and Leah Wasielewski and the HarperCollins marketing team; and, of course, Marissa Benedetto, Scooter Kaplan, Jen Rudin and Lisa Sharkcy.

And the women who volunteered their time (and bodies!) to be a part of our shoot for the reverses of the cover: Thank you so much!! I loved meeting all of you!!

TO THE PEOPLE WHO TAUGHT ME ABOUT MY MIND AND MY BODY, AND STARTED ME AND SUPPORTED ME ON THIS JOURNEY OF LEARNING . . .

Barry Michels for all your wisdom, guidance, expertise and tools, and for lending your knowledge to this project.

Master Cheung-yan Yuen, thank you for giving me one of the greatest gifts – the connection between my mind and my body. To Daxing Zhang and Tiger Hu Chen, for helping me to understand Master Cheung-yan Yuen's teachings and supporting me through the process.

Teddy Bass for helping keep my mind, body and spirit strong consistently over the last fourteen years. Through all the early mornings and changes in schedule, you've always helped me make sure that I get a workout in. Thank you for being my trainer, my partner in fitness, my friend.

And all of my teachers, mentors and partners in fun and fitness – you know who you are!

TO THE PEOPLE WHO HAVE BEEN THERE ALONG THE WAY – THANK YOU FOR THE SUPPORT AND INSPIRATION . . .

To ALL my girlfriends, thank you for all the glorious conversations about being a woman and for all the love and encouragement that I receive from you all and that you allow me to give to you. Your wisdom, insight, humour, honesty, strengths and weaknesses are what make life real. And I love all the different journeys I have taken with each and every one of you. I learn from you and you inspire me every day. I would especially like to thank my dear friend Elizabeth Berkley. Thank you for including me on one of your special field trips. The important work that you do with Ask Elizabeth inspired me to speak directly to women. Your inner beauty glows as brightly as your outer beauty.

To my sister, your strength and fortitude have given me inspiration for as long as I can remember. You're an amazing mother, daughter, wife and sister – all the hardest things to be. Thank you for all your support. And for always believing in me. You have always been at the core of everything I've ever done in my life.

C.E.E.C., you inspire me every day to want to put knowledge into the world, so that it can be a better place for you. I want you to have as much information as you can, because I want nothing more in this world than for you to be empowered to do the best and be the best that YOU can be. I love you with all my heart.

To all my family and friends, your love, support, laughter, care and eating skills keep the fire alight in my soul.

AND A GREAT BIG THANK-YOU TO ALL OF THE WOMEN WHO PICK THIS BOOK UP!

My parents instilled confidence in me. Part of why I wrote this book is because they always believed in me. And as a result of their confidence in me, I had confidence in myself. I in turn want to give that confidence to you, because the one thing that I always remembered my parents saying to me whenever I came up against any kind of challenge was that all I had to do was

MY best, not SOMEONE ELSE'S best – just MY OWN PERSONAL BEST. This way I was never in competition with anyone – the challenge wasn't to outdo someone else, because I could never be someone other than who I am, and that is good enough, no matter what other people are capable of.

Those words of support are what have allowed me to accomplish everything in my life, because there is NO SUCH THING AS FAILURE IF YOU ARE DOING THE BEST YOU CAN AT ANY GIVEN MOMENT UNDER ANY GIVEN CIRCUMSTANCE. But they also taught me that if I said I was doing my best and I wasn't, then I would know the truth, and it's better to be honest with yourself above anyone else, because you can't hide from yourself, and you always know if you're lying to yourself. So it's best just to make your best effort or admit to yourself that you could do better the next time around, and make the effort the next time that you have the opportunity. This is why I wrote this book: I want you to know what YOUR personal best is. And to know that if you do that at any given moment, then you are succeeding.

I hope that the information that this book holds will be a tool that will be useful to you whenever you are trying to do your best at taking care of your amazing body – and I thank you for letting me share with you.

21 That goes for kids, too: 'Childhood Obesity Facts', www.cdc.gov/healthyyouth/obesity/facts.htm, accessed 29 July, 2013.

21 the first generation: S. Jay Olshansky et al., 'A Potential Decline in Life Expectancy in the United States in the 21st Century', *New England Journal of Medicine*, 17 March, 2005, DOI: 10.1056/NEJMsr043743, accessed 29 July, 2013.

24 A Brief History of Food: The Food Timeline, www.foodtimeline.org, accessed 29 July, 2013.

25 salads that contain more than a thousand calories: '20 Salads Worse Than a Whopper', http://eatthis.menshealth.com/slideshow/print-list/186355, accessed 29 July, 2013.

38 they are still essential: 'Vitamins and Minerals', www.cdc.gov/nutrition/everyone/basics/vitamins, accessed 29 July, 2013.

41 The Whole Truth about Whole Grains: Jeannine Stein, 'The Whole Story on Whole Grains', *Los Angeles Times*, 31 May, 2010, http://articles.latimes.com/print/2010/may/31/health/la-he-whole-grains-20100531, accessed 29 July, 2013.

41 A seed has a few parts: Grain Foods Foundation, 'Whole Grain', Go with the Grain, www.gowiththegrain.org/nutrition/whole-grains.php, accessed 29 July, 2013.

44 100 grams of fibre: Robert H. Lustig, *Fat Chance: Beating the Odds against Sugar, Processed Food, Obesity, and Disease* (New York: Hudson Street Press, 2012).

45 laundry detergents: Mary Roach, *Gulp: Adventures on the Alimentary Canal* (New York: Norton, 2013).

47 at least ten times: Ibid.

49 when you ingest sucrose: Ibid.

50 5 pounds . . . 150 pounds (2 kilos . . . 68 kilos): Stephan Guyenet, 'By 2606, the US Diet Will Be 100 Per cent Sugar', Whole Health Source, 18 February, 2012, http://wholehealthsource.blogspot.com/2012/02/by-2606-us-diet-will-be-100-percent.html, accessed 29 July, 2013.

50 150 pounds (68 kilos) of sugar: 'How Much Sugar Do You Eat? You May Be Surprised!' www.dhhs.nh.gov/DPHS/nhp/adults/documents/sugar.pdf, accessed 29 July, 2013.

51 How Sugar Becomes Sugar: Robert L. Wolke, *What Einstein Told His Cook: Kitchen Science Explained* (New York: Norton, 2008).

52 chronic inflammation: Mark Hyman, 'Is Your Body Burning Up with Hidden Inflammation?' Huffpost Healthy Living, 27 August, 2009, www.huffingtonpost.com/dr-mark-hyman/is-your-body-burning-up-w_b_269717.html, accessed 29 July, 2013.

52 Being sedentary: Thomas Yates et al., 'Self-Reported Sitting Time and Markers of Inflammation, Insulin Resistance, and Adiposity', *American Journal of Preventive Medicine* 42, no. 1 (January 2012): 1–7.

54 variations to look out for: 'How to Spot Added Sugar on Food Labels', Harvard School of Public Health, www.hsph.harvard.edu/nutritionsource/added-sugar-on-food-labels/#1, accessed 29 July, 2013.

57 lean towards the sunlight: 'Protein Plays Role in Helping Plants See Light', Phys.org, http://phys.org/news/2011-10-protein-role.html, accessed 29 July, 2013.

57 The Amino Acid Trip: MedlinePlus, 'Amino Acids', www.nlm.nih.gov/medlineplus/ency/article/002222.htm, accessed 29 July, 2013.

59 How Much Is Enough?: Based on NHS RDA guidelines for protein intake for women older than nineteen years of age.

66 polyunsaturated fats: 'The Truth about Fats: Bad and Good', Harvard Medical School Family Health Guide, www.health.harvard.edu/fhg/updates/Truth-about-fats.shtml, accessed 29 July, 2013.

66 monounsaturated fats: Mayo Clinic staff, 'Dietary Fats: Know Which Types to Choose', www.mayoclinic.com/health/fat/NU00262, accessed 29 July, 2013.

67 coconut oil: Pina LoGiudice, 'The Surprising Benefits of Coconut Oil', *The Dr Oz Show*, www.doctoroz.com/videos/surprising-health-benefits-coconut-oil, accessed 29 July, 2013.

73 instead of a glass of milk: Laura Schocker, 'Surprisingly Calcium-Rich Foods That Aren't Milk', Huffington Post, 25 April 2012, www.huffingtonpost.com/2012/04/25/calcium-food-sources_n_1451010.html#slide=903353, accessed 29 July, 2013.

74 The Bone Builders: Institute of Medicine, *Dietary Reference Intakes for Calcium, Phosphorus, Magnesium, Vitamin D, and Fluoride* (Washington, DC: National Academies Press, 1997); and Institute of Medicine, *Dietary Reference Intakes for Calcium and Vitamin D* (Washington, DC: National Academies Press, 2011).

75 deficiency of D: '5 Tips for a Happier Life'. *The Dr Oz Show*, www.doctoroz.com/videos/5-tips-healthier-life, accessed 29 July, 2013.

76 Getting enough B_{12}: Kate Geagan, 'End Your Energy Crisis with Vitamin B_{12}', *The Dr Oz Show*, www.doctoroz.com/videos/end-your-energy-crisis-vitamin-b12?page=3#copy, accessed 29 July, 2013.

77 The Blood Formers: Institute of Medicine, *Dietary Reference Intakes for Thiamin, Riboflavin, Niacin, Vitamin B_6, Folate, Vitamin B_{12}, Pantothenic Acid, Biotin, and Choline* (Washington, DC: National Academies Press, 1998); and Institute of Medicine, *Dietary Reference Intakes for Vitamin A, Vitamin K, Arsenic, Boron, Chromium, Copper, Iodine, Iron, Manganese, Molybdenum, Nickel, Silicon, Vanadium, and Zinc* (Washington, DC: National Academies Press, 2001).

78 folic acid: Office on Women's Health, Department of Health and Human Services, 'Folic Acid Fact Sheet', http://womenshealth.gov/publications/our-publications/fact-sheet/folic-acid.cfm, accessed 29 July, 2013.

78 Free radicals: Jeanie Lerche Davis, 'How Antioxidants Work', WebMD, www.webmd.com/food-recipes/features/how-antioxidants-work1, accessed 29 July, 2013.

79 The Antioxidant Army: Institute of Medicine, *Dietary Reference Intakes for Vitamin C, Vitamin E, Selenium, and Carotenoids* (Washington, DC: National Academies Press, 2000); and Institute of Medicine, *Dietary Reference Intakes for Vitamin A, Vitamin K, Arsenic, Boron, Chromium, Copper, Iodine, Iron, Manganese, Molybdenum, Nickel, Silicon, Vanadium, and Zinc* (Washington, DC: National Academies Press, 2001).

80 Niacin helps: 'Vitamin B_3 (Niacin)', University of Maryland Medical Center, http://umm.edu/health/medical/altmed/supplement/vitamin-b3-niacin, accessed 29 July, 2013.

81 The Energy Vitamins: Institute of Medicine, *Dietary Reference Intakes for Thiamin, Riboflavin, Niacin, Vitamin B_6, Folate, Vitamin B_{12}, Pantothenic Acid, Biotin, and Choline* (Washington, DC: National Academies Press, 1998).

83 The Hydrating Electrolytes: Institute of Medicine, *Dietary Reference Intakes for Water, Potassium, Sodium, Chloride, and Sulfate* (Washington, DC: National Academies Press, 2005).

84 indoles: Susan C. Tilton, 'Benefits and Risks of Supplementation with Indole Phytochemicals', Linus Pauling Institute, *Research Newsletter*, Spring–Summer 2006, http://lpi.oregonstate.edu/ss06/indole.html, accessed July 30, 2013.

91 flavour hit: Mary Roach, *Gulp*.

94 ATP: 'Cellular Respiration', Department of Biology, Indiana University–Purdue University Indianapolis, www.biology.iupui.edu/biocourses/N100/2k4ch7respirationnotes.html, accessed 1 August, 2013.

96 fart: Mary Roach, *Gulp*.

100 bacterial cells: Carl Zimmer, 'How Microbes Defend and Define Us', *New York Times*, 12 July 2010, www.nytimes.com/2010/07/13/science/13micro.html?_r=2&pagewanted=all, accessed 2 August, 2013.

100 pencil eraser: 'Bacterial Infection', MedlinePlus, www.nlm.nih.gov/medlineplus/bacterialinfections.html, accessed 2 August, 2013.

100 that colonization: Nathan Wolfe, 'Small, Small World', *National Geographic*, January 2013, http://ngm.nationalgeographic.com/2013/01/125-microbes/wolfe-text, accessed 2 August, 2013.

101 two and a half: Antonio González and Yoshiki Vázquez-Baeza, 'The Assembly of an Infant Gut Microbiome Framed against Healthy Human Adults', Knightlab, University of Colorado–Boulder, www.youtube.com/watch?v=Pb272zsixSQ, accessed 2 August, 2013.

102 Some bacteria: Mozzi Fernanda et al., *Biotechnology of Lactic Acid Bacteria* (Hoboken, NJ: Wiley-Blackwell, 2010).

102 Cows on Penicillin: 'Beef Procedures: Antibiotic Use', South Dakota State University Veterinary Extension, www.sdstate.edu/vs/extension/beef-procedures-antibiotics.cfm, accessed 29 July, 2013.

102 scientists are researching: Conversation with Dr Martin Blaser, director of the Human Microbiome Program at NYU, 19 February, 2013.

103 *Bifidobacterium infantis*: 'Supplements for IBS: What Works?' Irritable Bowel Syndrome (IBS) Health Center, WebMD, www.webmd.com/ibs/features/supplements-for-ibs-what-works, accessed 29 July, 2013.

104 Metchnikoff: Thomas J. Montville and Karl R. Matthew, *Food Microbiology: An Introduction* (Washington, DC: American Society for Microbiology, 2008).

105 *Lactobacillus delbrueckii bulgaricus* and *Streptococcus thermophilus*: Ibid.

105 fifty billion active *L. acidophilus*: 'Probiotics FAQ', Bio-K Plus, www.biokplus.com/en-us/about-probiotics/probiotics-faq#19n3585, accessed 29 July, 2013.

112 on the job: Timothy S. Church et al., 'Trends over 5 Decades in U.S. Occupation-Related Physical Activity and Their Associations with Obesity', *PLOS One*, 25 May, 2011, www.plosone.org/article/info%3Adoi%2F10.1371%2Fjournal.pone.0019657#s1.

112 taking care of a home: Edward Archer et al., '45-Year Trends in Women's Use of Time and Household Management Energy Expenditure', www.plosone.org/article/info%3Adoi%2F10.1371%2Fjournal.pone.0056620#ack, accessed 30 July, 2012.

118 Benefits of Exertion: Alyssa Shaffer, 'Power Surge: The Hidden Benefits of Exercise', *Fitness*, www.fitnessmagazine.com/workout/motivation/get-started/power-surge-the-hidden-benefits-of-exercise, accessed 2 August, 2013; and Gretchen Reynolds, 'Modera-

tion as the Sweet Spot for Exercise', Well, *New York Times*, 6 June, 2012, http://well.blogs.nytimes.com/2012/06/06/moderation-as-the-sweet-spot-for-exercise, accessed 2 August, 2013.

125 Obesity and diabetes: Jerry N. Morris et al., 'Incidence and Prediction of Ischaemic Heart Disease in London Busmen', *Lancet* 288, no. 7463 (10 September, 1966): 553–59; Frank B. Hu et al., 'Television Watching and Other Sedentary Behaviors in Relation to Risk of Obesity and Type 2 Diabetes Mellitus in Women', *Journal of the American Medical Association* 289, no. 14 (9 April, 2003): 1785–91; Frank B. Hu et al., 'Physical Activity and Television Watching in Relation to Risk for Type 2 Diabetes Mellitus in Men', *Archives of Internal Medicine* 161, no. 12 (25 June, 2001): 1542–48; and David W. Dunstan, Bethany Howard, Genevieve N. Healy, and Neville Owen, 'Too Much Sitting – a Health Hazard', *Diabetes Research and Clinical Practice* 97, no. 3 (September, 2012): 368–76, 2013.

126 increased risk factors for diabetes: David W. Dunstan et al., 'Breaking Up Prolonged Sitting Reduces Postprandial Glucose and Insulin Responses', *Diabetes Care* 35, no. 5 (May 2012): 976–83.

127 Everybody Has Ten Minutes: Carol Ewing Garber et al., 'Quantity and Quality of Exercise for Developing and Maintaining Cardiorespiratory, Musculoskeletal, and Neuromotor Fitness in Apparently Healthy Adults: Guidance for Prescribing Exercise'. American College of Sports Medicine Position Stand, *Medicine & Science in Sports & Exercise* 43, no. 7 (July 2011): 1334–59.

128 When you train and lose weight: E. V. Menshikova, 'Characteristics of Skeletal Muscle Mitochondrial Biogenesis Induced by Moderate-Intensity Exercise and Weight Loss in Obesity', *Journal of Applied Physiology* 103, no. 1 (July 2007): 21–27, www.ncbi.nlm.nih.gov/pubmed/17332268.

144 Myocardial cells: Charles R. Morris, *The Surgeons: Life and Death in a Top Heart Center* (New York: Norton, 2007).

144 Oxygen is brain food: Maria Chiara Gallotta, 'Effects of Varying Types of Exertion on Children's Attention Capacity', *Medicine & Science in Sports & Exercise* 44, no. 3 (March 2012): 550–55.

148 Bone by Bone: 'Human Body & Mind', BBC Science, www.bbc.co.uk/science/humanbody, accessed 2 August, 2013.

154 cardiac muscles: Morris, *Surgeons*.

180 apoptosis: Natalie Angier, *Woman: An Intimate Geography* (New York: Anchor, 2000).

184 female athlete triad: Aurelia Nattiv et al., 'The American College of Sports Medicine Position Stand on the Female Athlete Triad', *Medicine & Science in Sports & Exercise* 39, no. 10 (October 2007): 1867–82, http://journals.lww.com/acsm-msse/Fulltext/2007/10000/The_Female_Athlete_Triad.26.aspx, accessed 24 August, 2013.

184 carb-crazy: Judith Wurtman, 'You Can Prevent PMS from Destroying Your Diet', The Antidepressant Diet, *Psychology Today* blog, www.psychologytoday.com/blog/the-antidepressant-diet/201008/you-can-prevent-pms-destroying-your-diet, accessed 30 July, 2013.

186 allergies, asthma: 'C-Section May Raise Child's Risk of Allergies, Asthma: Study', *U.S. News & World Report*, February 25, 2013, http://health.usnews.com/health-news/news/articles/2013/02/25/c-section-may-raise-childs-risk-of-allergies-asthma-study, accessed 30 July, 2013.

186 childhood obesity: Genevra Pittman, 'Babies Born via C-Section Linked to Child Obesity', Reuters, 24 May 2013, http://news.msn.com/science-technology/babies-born-via-c-sections-linked-to-child-obesity, accessed 30 July, 2013.

188 tired banker: Sam Ashton, 'Dozy Banker Sleeps on Keyboard, Transfers £190M', 4 June 2013, MSN Money, http://money.uk.msn.com/trending-blog/dozy-banker-sleeps-on -keyboard-transfers-%C2%A3190m, accessed 30 July, 2013.

189 studies of medical students: DeWitt C. Baldwin Jr. and Steven R. Daugherty, 'Sleep Deprivation and Fatigue in Residency Training: Results of a National Survey of First- and Second-Year Residents', *Sleep* 27, no. 2 (2004), www.journalsleep.org/ViewAbstract .aspx?pid=25943, accessed 30 July, 2013.

195 Your sleep: 'Why Do We Sleep, Anyway!' Healthy Sleep, Harvard Medical School, http:// healthysleep.med.harvard.edu/healthy/matters/benefits-of-sleep/why-do-we-sleep, accessed 30 July, 2013.

223 This Is Your Brain on Food: David Kessler, *The End of Overeating: Taking Control of the Insatiable American Appetite* (New York: Rodale, 2009).

231 Habit Loop: Charles Duhigg, *The Power of Habit: Why We Do What We Do in Life and Business* (Random House: New York, 2012).

SUPPLEMENTARY READING LIST

Want to know more? Me too. You might want to check out:

Angier, Natalie. *Woman: An Intimate Geography*. New York: Houghton Mifflin Harcourt, 1999.

Ariely, Dan. *Predictably Irrational: The Hidden Forces That Shape Our Decisions*. New York: Harper Perennial, 2010.

Baumeister, Roy F. and John Tierney. *Willpower: Rediscovering the Greatest Human Strength*. New York: Penguin, 2012.

Boston Women's Health Book Collective and Judy Norsigian. *Our Bodies Ourselves*. New York: Simon & Schuster, 2011.

Duhigg, Charles. *The Power of Habit: Why We Do What We Do in Life and Business*. New York: Random House, 2012.

Fallon, Sally and Mary Enig. *Nourishing Traditions: The Cookbook That Challenges Politically Correct Nutrition and the Diet Dictocrats*. Warsaw, Ind.: New Trends Publishing, 1999.

Kahneman, Daniel. *Thinking, Fast and Slow*. New York: Farrar, Straus and Giroux, 2011.

Kessler, David A. *The End of Overeating: Taking Control of the Insatiable American Appetite*. New York: Rodale, 2009.

Lovegren, Sylvia. *Fashionable Food: Seven Decades of Food Fads*. Chicago: University of Chicago Press, 2005.

Lustig, Robert H. *Fat Chance: Beating the Odds Against Sugar, Processed Food, Obesity, and Disease*. New York: Hudson Street Press, 2012.

Pollan, Michael. *The Omnivore's Dilemma: A Natural History of Four Meals*. New York: The Penguin Press, 2006.

Roach, Mary. *Gulp: Adventures on the Alimentary Canal*. New York: W. W. Norton & Company, 2013.

Thaler, Richard H. and Cass R. Sunstein. *Nudge: Improving Decisions About Health, Wealth, and Happiness*. New Haven, Conn.: Yale University Press, 2008.

Wolcke, Robert L. *What Einstein Told His Cook: Kitchen Science Explained*. New York: W. W. Norton & Company, 2008.

Wansink, Brian. *Mindless Eating: Why We Eat More Than We Think*. New York: Bantam, 2007.

www.foodtimeline.org.

INDEX